The Urban Book Series

The Urban Book Series is a resource for urban studies and geography research worldwide. It provides a unique and innovative resource for the latest developments in the field, nurturing a comprehensive and encompassing publication venue for urban studies, urban geography, planning and regional development.

The series publishes peer-reviewed volumes related to urbanization, sustainability, urban environments, sustainable urbanism, governance, globalization, urban and sustainable development, spatial and area studies, urban management, transport systems, urban infrastructure, urban dynamics, green cities and urban landscapes. It also invites research which documents urbanization processes and urban dynamics on a national, regional and local level, welcoming case studies, as well as comparative and applied research.

The series will appeal to urbanists, geographers, planners, engineers, architects, policy makers, and to all of those interested in a wide-ranging overview of contemporary urban studies and innovations in the field. It accepts monographs, edited volumes and textbooks.

Now Indexed by Scopus!

More information about this series at http://www.springer.com/series/14773

Shlomit Flint Ashery

Spatial Behavior in Haredi Jewish Communities in Great Britain

Springer

Shlomit Flint Ashery
Department of Geography and Environment
Bar Ilan University
Ramat Gan, Israel

ISSN 2365-757X ISSN 2365-7588 (electronic)
The Urban Book Series
ISBN 978-3-030-25857-3 ISBN 978-3-030-25858-0 (eBook)
https://doi.org/10.1007/978-3-030-25858-0

This Springer imprint is published by the registered company Springer Nature Switzerland AG
The registered company address is: Gewerbestrasse 11, 6330 Cham, Switzerland

This book is dedicated to the memory of my parents, Tali and Itzhak Flint

Contents

List of Figures

List of Tables

Chapter 1
Introduction

Abstract Orthodox Judaism is deeply spatial. It requires that its members live in accordance with Halakha (Jewish law), the dominant protocol determining Jewish ritual and translation of the commandments into many aspects of day-to-day living. The strict orthodox Jewish (Haredi) community comprises many sects whose communal identity plays a central role in everyday life and spatial organisation. This book examines how different levels of communal organisation are reflected in the residential patterns of four British communities: The Litvish communities of Golders Green and Gateshead, and the Hassidic community of Stamford Hill and Canvey Island. The research reveals and analyses powerful mechanisms of residential segregation acting at the level of the apartment, building and the near neighbourhood level. Identifying the main engines of spontaneous and organised neighbourhood change and evaluating the difficulties of liberal planning dealing with non-autonomous individuals in the housing market shed light on similar processes occurring in other city centres with diverse population groups. This book aims to highlight the impact of various organisational levels on the spatial structure of the urban enclave. It focuses on the internal dynamics of ethno-religious enclaves that emerge from three levels of action: (1) individuals' relationships with their own and other groups; (2) the community leadership's powers inside the group and in respect of other groups; and (3) government directives and tools (e.g. planning). The study investigates the residential dynamics of four Haredi areas in the UK, representing differing social organisational through communal leadership and local-government directives.

Keywords Residential behaviour · Orthodox Judaism · Enclave · Liberal planning · Borders · Delineation · British Haredi Jewry

Orthodox Judaism is deeply spatial. It requires that its members live in accordance with Halakha (Jewish law), the dominant protocol determining Jewish ritual and translation of the commandments into many aspects of day-to-day living (Shilhav and Friedman 1985). It composed of groups and subgroups (sects), and the residential preferences of the Haredi (Strictly orthodox Jews) individuals are profoundly affected by the need to live in a strict religious environment. Most of the Haredi belong to the main groups: The Litvish, the Hasidim and the Sephardic. All wish to live among "friends", other members of the same group to which they belong, but

S. Flint Ashery, *Spatial Behavior in Haredi Jewish Communities in Great Britain*, The Urban Book Series, https://doi.org/10.1007/978-3-030-25858-0_1

differ in terms of their self-identity and their status within Haredi society. Over the years, lifestyle characteristics, including limited career training and the conventional devotion to Torah study, together with the innate conflict with modern lifestyles, have motivated voluntary territorial separation of the Haredi population into an enclave of sort. The objective of the Haredi struggle for territorial segregation is the wish to safeguard the younger generation from the perceived threats of modern culture. In the recent decade, however, with the growing pressure on the traditional enclaves, the Haredi gradually transforms from an "enclave culture" into more "assertive culture" and emphasises on accomplishing territory through the use of spatial competition. Although this enigmatic population tends to form sizeable enclaves in large cities around the globe, very little has been written about their socio-spatial behaviour.

High population density and increased prices for flats in Haredi enclaves have motivated a constant migration of population from the traditional enclaves. This migration process increases the direct and ongoing contact within and between the Haredi communities themselves, as well as the everlasting rift between tradition and modernism. While the traditional world sanctified the established social order and emphasised living inside the community, from the cradle to the grave, as a basis for individual's definition in society, modern society highlights the individual's qualities and preferences as the means for defining one's identity. In modern society, religious tradition is but one of many other ideological possibilities, struggling for its place in a diverse marketplace of values (Flint Ashery 2018a). "Haredi-fied" neighbourhoods have two key characteristics: proximity to an existing Haredi congregation and location close to inner-city neighbourhoods (Alfasi et al. 2013). Experience in Israel and the USA revealed that the influx of Haredi into non-Haredi neighbourhoods had caused friction and bitter struggles over the city's character. These local territorial communities are not merely subservient to the influence of central authorities: They require a local solution from the municipal planning authorities and sometimes have a considerable impact on regimes (Kolossov 2005; Flint 2015; Newman 2016). Although our understanding of the role played by local communities in determining our multi-layered territorial compartments is less rigid than in the past, the impact of these interest groups on planning is as yet it remains undertheorised.

In recent decades, with the dissolution of the social, economic and political frameworks that constitute the background for planning, the impact and the pressure of direct cooperation of interest groups on urban space had considerably increased (Davies 2001; Alexander 2002; Kolossov 2005). "Interests" are defined as motivations for action which affect social behaviour, judgment and choices, and the aim of an "interest group" is to influence public opinion and/or policy (Miller 1999). The need to protect the public interest and guarantee its spatial rights have led to the establishment of a hierarchical system intervenes in local planning policy throughout the developed world. Influenced by distinct local culture and norms, the liberal mechanism needs to identify the various communities and balance their interests in accordance with local needs (Dean 2011). The innate tension between those worldviews and the planning system's difficulty in striking a balance between the various interests stir up a range of political and spatial problems (Campbell 2002; Yiftachel 2006).

In the liberal ideology, it is only in benighted, backward or premodern societies that individuals put an unchosen group identity—such as membership of an ethnic group or nation—ahead of their interests as individuals. Thus, spatial pattern of ghettoisation, segregation and sectarianism that obscure the fundamental interests which humans have in common is considered as outsiders to the civilised liberal world (Wastl-Walter 2016). The British planning system lies on the seam between charting national policy and the pressures of diversified urban politics, subjected to the regulatory and hierarchy planning system. Planning, in this respect, thinks in terms of individuals rather than collectivities, seeks for the erosion of ethnic solidarities and weakens the boundaries between groups. Planners wish for weak boundaries in order to integrate specific urban qualities (e.g. low boundaries towards diverse populations or a rich variety of lifestyles, cuisine and creativity). Recently, with the growing impact of interest groups on urban space, the assumption of liberal planning that democratic institutions, including the market, will "balance" competing interests in accordance with local needs has been weakened (Flint Ashery 2017; Murray 2017). In urban reality, the main key actors in the planning process, planners and politicians, had to cope with interest groups characterised by diverse institutional structures, access to resources, holding inconsistent territorial interests (Taylor 1998). The tolerance of systems towards individuals from private investors and strong interest groups—mainly religious groups—involved in planning increased. This process emphasises the difficulties of the individualistic-orientated institutes to deal with the residential behaviour of groups. A particular challenge to the liberal planning systems poses groups that are committed to non-liberal laws and concepts, like the Haredi Jews.

In the UK, the Haredi are a growing minority within the British "Mainstream" Jewry. This research contends that in order to examine residential behaviour and processes within these communities, one must refer to the fundamental social structures and values from which affected communities draw their strength. The relations between the Haredi groups examined in this study reflect those within the entire Haredi world, while local circumstances affect the specific links connecting the subgroups. Through an analysis of the developments and characteristics of four representative residential areas in Golders Green, Stamford Hill, Gateshead and Canvey Island, the research will, therefore, examine how interest groups characterised by different levels of cooperation between the individual and his group affect the residential structure of the urban environment, and will discuss the difficulties of liberal planning to implement its agenda and regulate resources between groups and individuals. In order to explain the way individuals inclined towards and cooperate with groups, the research will use the themes of interest groups and group behaviour in the urban space. The theme of nested residential patterns, territoriality, microsegregation and reshaping neighbourhoods (Flint Ashery 2018b) will elaborate the way different groups adopt different strategies to claim space.

This work is both methodological and empirical. The methodological objective is to propose complementary qualitative and quantitative procedures to improve understanding of residential processes and pattern considering the underenumeration of the Haredi population. Due to the different nature of the communities and the variety of research areas, the test cases cannot be compared with a general methodology that suits them all. Therefore, an appropriate methodology was constructed for each case study, which can provide us with the ability to analyse the area at the highest resolution possible. High-resolution examination of these residential tendencies will describe the rich potential of bottom-up research design and suggest its applicability to other social phenomena. The empirical objective of this study is to compare the mechanisms used by members of the Litvish and the Hasidic communities in four residential areas in the UK to construct a private and public "demarcation line" between the abstract obligations and the spatial reality in which they are applied. Through this, the work aims to contribute to wider discussions about identity, segregation and place and to add to a growing literature of "new" theoretically informed geographies of religion (Kong 2001; Holloway and Valins 2002; Newman 2016). The study begins with a description of the socio-spatial delineation of the Haredi communities and the development of the "Jewish North-Western route" before explaining, in Chap. 3, the difficulties in enumerating Haredi Jews in the UK. Chapter 4 traces the links between social norms and the urban environment and reveals the significant mechanisms that determine the Litvish community's residential behaviour. It describes both the formation of nested residential patterns of Golders Green and the reshaping of the neighbourhood's territoriality in Gateshead. Chapter 8 focuses on the effects of communal identity on intergroup residential relations within the Hasidic world. Chapter 9 presents the non-economic segregation, which provides the context for the Hasidic residential behaviour in Stamford Hill and Chap. 10 describes the group action in Canvey Island as a driving force for neighbourhood change. To ensure anonymity and confidentiality, pseudonyms were used, and all potential identifiers were removed or modified in the interviews' transcripts. Through an analysis of these neighbourhoods' development and characteristics, Chap. 11 discusses the impact of intra-communal and intergroup relations within the Hasidic world on the spatial organisation and traces the links between social norms and the urban environment.

References

Alexander ER (2002) The public interest in planning: from legitimation to substantive plan-evaluation. Plan Theory 1:226–249

Alfasi N, Flint Ashery S, Benenson I (2013) Between the individual and the community: residential patterns of the Haredi population in Jerusalem. Int J Urban Reg Res 37(6):2152–2176

Campbell H (2002) Planning: an idea of value (Inaugural Lecture). Town Plan Rev 73(3):271–288

Davies AR (2001) Hidden or hiding? Public perceptions of participation in the planning system. Town Plan Rev 72:193–216

Dean H (2011) Welfare, identity and the life course. In: Baldock J, Mitton L, Manning N, Vickerstaff S (eds) Social policy. Oxford University Press, Oxford, UK, pp 201–244. ISBN 9780199570843

Flint Ashery S (2015) Public welfare or sectarianism: a new challenge for planning. Plan Theory & Pract 16(3):299–318

Flint Ashery S (2017) 'Decision not to decide': a new challenge for planning. Eur Plan Stud 25(6):1076–1098

Flint Ashery S (2018a) Schelling-type micro-segregation in a Hassidic enclave of Stamford-Hill. Hous Stud 33(7):1038–1059

Flint Ashery S (2018b) Horizontal examination: micro-segregation mechanism in the diverse area of Whitechapel. In: Micro-residential dynamics, pp 25–34. Springer, Cham

Holloway J, Valins O (2002) Placing religion and spirituality in geography

Kolossov V (2005) Border studies: changing perspectives and theoretical approaches. Geopolitics 10(3):606–632

Kong L (2001) Mapping 'new' geographies of religion: politics and poetics in modernity. Prog Hum Geogr 25(2):211–233

Miller D (1999) The norm of self-interest. Am Psychol 54(12):1053–1060

Murray MJ (2017) The urbanism of exception. Cambridge University Press

Newman D (2016) Contemporary research agendas in border studies: an overview. In: The Routledge research companion to border studies. Routledge, pp 55–70

Shilhav Y, Friedman M (1985) Growth and segregation: the Haredi community of Jerusalem. The Jerusalem Institute for Israel Studies, Jerusalem (in Hebrew)

Taylor N (1998) Urban planning theory since 1945. Sage Publications, London, U.K

Wastl-Walter D (ed) (2016) The Routledge research companion to border studies. Routledge

Yiftachel O (2006) Ethnocracy: land and identity politics in Israel/Palestine. University of Pennsylvania Press, Philadelphia

Chapter 2
The Haredi Jews in the UK

Abstract Since the seventeenth century, Jews in the UK have lived predominately in certain areas. Following the Second World War and the annihilation of most of Eastern European Jewry, the organisation of Haredi society worldwide underwent dramatic change. Holocaust survivors, along with Jewish immigrants from Islamic countries, moved to major urban centres in the West and Israel, and the British Jewry community has expanded. The Jews in the UK had been gradually accepted both politically and socially into almost every sphere of activity and had followed the patterns of the wider host community. Most integrated slowly into the broader modern society, adopting different combinations of religious and ethnic lifestyle, education and employment, and the character of the Jewish life in Britain—their need to live among people of their group—has changed. To an outside observer, Anglo-Jewry is the product of British society. Today, the British Jewish community, the case study of this study, contains two main subcommunities: The Haredi and the mainstream. There is considerable heterogeneity among these subcommunities and a blurry boundary between them. The Haredi community is by far the fastest-growing component of contemporary British Jewry and Jewry worldwide. In its orthodox interpretation, a hierarchical spiritual leadership, which mediates between the religious commandments and their interpretation and implementation in practice, interweaves the life of the Haredi individual with a series of communal commitments whose fulfilment is not open to debate. The Haredi community, exhibit visually and locationally distinct characteristics, comprise an increasing proportion of the Jewish population of the UK.

Keywords Delineation of the Haredi identity · Haredi sub-communities · The British Jewish communities · The Jewish North-Western route · Orthodox Judaism · British Jewish and the general society

2.1 Delineation of the Haredi Identity

In its orthodox interpretation, a hierarchical spiritual leadership, which mediates between the religious commandments and their interpretation and implementation in practice, interweaves the life of the Haredi individual with a series of communal

S. Flint Ashery, *Spatial Behavior in Haredi Jewish Communities in Great Britain*,
The Urban Book Series, https://doi.org/10.1007/978-3-030-25858-0_2

commitments whose fulfilment is not open to debate. The accepted definition of a Haredi family is "a family in which at least one member studied, or is studying, at a Yeshiva which is not a high-school Yeshiva" (Dahan 1999). This definition includes a great many diverse communities which, along with punctilious observance of Jewish law, are also characterised by cultural conservatism. However, the Haredi is bound by a shared faith and culture and, as Valins (2000) describes are "known for their traditional style of dress, their stringent interpretations of ways of life inscribed in ancient, sacred texts and their explicit construction of socio-spatial boundaries to separate themselves from those they consider to be 'other'".

Following the Second World War and the annihilation of most of Eastern European Jewry, the organisation of Haredi society worldwide underwent dramatic change. Holocaust survivors, along with Jewish immigrants from Islamic countries, moved to major urban centres in the West and Israel. Most integrated into the workforce of the broader modern society, adopting different combinations of religious and ethnic lifestyle, education, and employment. Aspinall (2000) differentiates between belonging to a religious group, which means sharing a "common set of beliefs" (Aspinall 2000:587), and belonging to an ethnic group, which means "shared characteristics of ancestry and culture" (Aspinall 2000:587). He further suggests that "it is difficult to differentiate between religion and ethnicity" in practice especially as there is a lot more to Judaism than just religion or just ethnicity. Aspinall considered that in Judaism, there is a continuum between ethnicity and religion as observing religious practices is more than religion but as a means of identifying with the Jewish community (Aspinall 2000:591).

Previous studies explain that similar to other religious minority groups (Finke 1997, 2015; Phillips 2006; Tausch et al. 2011; Flint 2018b), the Haredi are characterised with strong solidarity bonds and an intense sense of spatial belonging and urban identity. The Haredi family is exceptional in its awareness of its own and neighbours' biography (Flint et al. 2012; Alfasi et al. 2013). Neighbours' identity and a flat's genealogy are important components of the stated preferences upon making residential choices. Different values, interpretations of religious ritual and normative behaviour motivated the individuals to concentrate with other members of the same subgroup to which they belong and create friendly social environments. Expression of this distinctiveness is found in the clear tendency to avoid the larger Jewish society, the strong discourage time wasting on anything not directly related to Torah, and the need to live among other followers of Haredi. The need to concentrate with neighbours who share the same faith, worldview and lifestyle directed them towards living in a sufficient circle of "friends" that matches up with the person's self-identity, and within a short distance from their communities' synagogues. Berman (2000) explains that people "engage in a strict practice of ritual and maintain the lifestyles reflecting their national origins". Belonging to the group's territory is advantageous regarding community services, mutual assistance and support, as well as the socialisation process of the younger generation which is kept isolated from external influences. A territorial concentration of group's members facilitates the leadership's control of members' lives, thus preventing deviation from accepted norms" (see also Shilhav and Friedman 1985). Haredi enclaves are characterised by apparent dominance in the public space and clear territorial boundaries.

The role of economic considerations is not absent, though. Shilhav and Friedman (1985) explain how lifestyle characteristics, including limited career training, high fertility rates and the traditional devotion to Torah study, prevent economic improvement for the majority of Haredi Jewish households (see also Gonen 2006). Haredi households have large variations in their financial positions. In the UK, a gradually growing minority of the Haredi population is of comfortable means, but over 50% are on means-tested benefits (Interlink 2010). Haredi communities, however, are sometimes well off: many Haredi communities control an array of financial resources and services donated by community members located throughout the world. These funds finance an extensive system of service provision attuned to individual and family needs in the areas of education, marriage (including apartment purchases for newlyweds) and welfare (Berman 2000; Blumen 2007). The operation of this unique instrument for isolating the community and supervising its members emphasises the role of individual's dependence on the community and its leaders in the formation of the spatial forms (Yiftachel et al. 2016). For this research, the tendency of people of different economic circumstances to live side by side enables us to minimise economic variables in the high-resolution examination of Haredi residential dynamics and to focus on the role of identity and the importance of living among friends in the creation of residential patterns.

This sociocultural tendency, together with the innate conflict with modern, secular lifestyles, has motivated, during the years, a voluntary territorial separation of the Haredi population into a ghetto of sorts (Shilhav 1993; Chakak 2004). The spatial division allows their religiously based lives to remain cut off from modern, especially secular life (Ben-Sasson 2008). Despite the Haredi willingness to accept elements of modernity so long as they are not perceived as a threat to its cultural values, the desire to isolate the younger generation from different values demands a rejection of the value-related elements of modernity. Sivan and Caplan (2003) point out that the "Jewish Haredi society is very selective in what aspects of modernity it accepts: high percentages of this population use cellular phones (about 87%) although their level of motorisation is low and their use of elevators in residential buildings is rare. This phenomenon has had a direct effect on building height and neighbourhood compactness, which remains tied to their self-imposed ghettoisation despite the inconveniences flowing from the overcrowding and poor standard of living in those same neighbourhoods". The retreat from external influences into the Haredi enclave turns the urban space into a sought-after resource, allowing the leadership to consolidate the community and to oversee the formation of group identity within the defined Haredi area.

2.2 The Haredi Population Groups

While the Haredi appears unified to outsiders, it is distinguished by internal, nuanced distinctions among the Haredi groups and subgroups. Those internal divisions are affecting the values, religious rituals and normative behaviour and vary from one subgroup to another regarding clothing, language and attitudes towards working as opposed to full-time Torah (Bible) study. The definition of the individual's role in the Haredi society is determined according to his genealogical affiliation with one of three main groups, named after the communal origin: Litvish (also known as "Litvak" or Lithuanian or Misnagdim—"opponents" of the Hasidim), Hasidic or Sephardic (Shilhav 1993). The groups' labels reflect internally generated distinctions. The three groups mentioned above are quite similar in their wish to maintain a high degree of voluntary residential segregation within a strict religious environment. However, they are distinguished from one another regarding their status within Haredi society, language and custom and approach to employment as opposed to engaging in the religious study (Dahan 1999; Lupo 2004; Shilhav and Friedman 1985). Their attire particularly distinguishes the Hasidim, usually consisting of black suits and hats, the men usually sporting beards. Note that while theological distinctions are beyond the scope of this study, the research provides some distinguishing features to support our findings in the areas of the subgroups' socio-spatial behaviour and interactions.

The roots of Haredi are in Eastern Europe, which is where the movement was established. The Jews from these countries are called Ashkenazi. Even though over the last few decades, there emerged Haredi communities comprised of Jews of Middle Eastern descent (i.e. Sephardim) the Haredi movement is still Ashkenazi in many respects. In the Haredi world, the basic lifestyle and ritual orientation of each subgroup express theological stances and historical tradition determined by their association to the group's leader, and named after the geographic origin (Shilhav 1993). The historical dispersion of these communities among different geographical regions led to the adoption of different cultural codes, sometimes even leading to cultural distinctions among different sectors of the same community. In this respect, Vaughan and Penn (2006) state that: "The incidence of endogamy amongst Jewish immigrants was found to be significantly higher than that of gentile immigrants and was not simply a case of Jews marrying other Jews, but the marriage was tied back very clearly to the country of origin. This tendency has been explained by Godley (1997) who shows that, in many cases, single young males would send back home for a bride once they had established themselves financially". The distinctions among the groups are widely acknowledged among the Haredi who affiliate the individuals accordingly, and the segregated attitudes expressed by the individuals (Alderman 2011). Aran (2013) argues that although the Hasidim and the Litvish "do not necessarily differ regarding the relatively high degrees of their religiosity, the present rivalry between these two camps is not as intense as in the past". The Litvish represent the Haredi elite, setting the norms for the whole Haredi society, aspiring to rule adherents' minds rather than their emotions and putting religious studies as the core value of an individual's life (Etkes and Tikochinski 2004). Litvish groups are differentiated by their connection to a specific rabbi (religious leader) or Yeshiva (institution of higher religious learning). The Litvish ascribe holy studies as a central cultural component in everyday's life (Gonen 2006). Men's lives are therefore centred on the Yeshivas, a higher religious studies institution. The Litvish nonetheless wear modern black suits and use English as their daily language (Etkes and Tikochinski 2004; Zelkin 2006). The various Hasidic groups, however, emphasise an emotional religious fervour based on awe of God. Their lives focus on the Admor's (the religious leader) court, which fulfils both religious and active leadership functions in the community's daily life (Green 2001; Piekarz 1999). Hasidim maintain most of their social relationships within a specific Hasidic court, including marriage (Ben-Sasson 2008).

These two traditions, the focus of this book, are, as Aran (2013) describes, "still distinguishable by their different lifestyles, worship and leadership models, social and economic institutions, and separate political organisations". Each one of these groups follows a slightly different approach to following the regulations in the sacred texts, interested in guarding its unique cultural identity and can be identified by variations in their style of dress. Valins (2003a, b) explains that "each of these groups is further subdivided, so that within Hasidim, for example, there are numerous groups each led by a religious leader (a rabbi). The major Hasidic groups include Belz, Bobov, Ger, Lubavitch and Satmar (with their names typically derived from their town of origin). To identify with particular ethnic, cultural, religious or other groupings is a crucial component of human behaviour. Also, these Jews frequently construct spatial devices known as "Eruv", which are complex devices consisting of poles and wires that surround communities. Eruv symbolically changes the classification of public or karmelis (sometimes described as "semi-public") space into private ownership. By reclassifying space, restrictions on the "carrying" (in a Jewish legalistic sense) of items such as walking sticks, wheelchairs and prams are removed, thus giving freedom to people who would otherwise be trapped in their homes over the Sabbath". Valins (2003a, b) also explains that the "political and legalistic challenges of constructing these devices are incredibly high" (see also Bechhofer 1998) and provides a clear example of how certain Jews seek to bound particular spatial areas (Valins 2000; Vincent and Warf 2002).

2.3 The British Jewish Communities

Since 1656, when a small colony of some 100 Sephardic Jews living in London came into the open again and allowed to remain, Jews have lived predominately in certain areas (Newman 1976). One hundred ten people lived within the walls of the small parish of London. They formed about 10% of the total population of these parishes. Between 1800 and 1870, the Jewish population was doubled and reached up to 40k. The establishment of these new centres has raised the need for the provision of new centres of worship, to enable to worship in a synagogue near enough for the members to walk to on a Sabbath morning. Six main synagogues in London had served the population of the East side of the city: Duke's place, Fenchurch Street,

Bishopsgate and Finsbury Square, towards Goodman's Fields and Houndsditch in addition to some small and significant synagogues. Since then, the key to much of Anglo-Jewish history as of English history in the nineteenth century is the growth of the middle class.

The influx of new Jewish migrants to London during the 1880s increased the consciousness of the division between the West End's rich Anglo-Jewish, and the poor in the East End. In the decade follows, the terms "East End" or "Whitechapel" became synonyms to a different world, poor and overcrowded, within the same city. Up until the 1950s, Jews formed the only significant non-Western-European and non-Christian minority group in Britain. The leadership was in the hands of the old families who wished to rationalise the synagogue system and to provide for future growth. As Sapiro (2016) describes, "Immigration from a range of world regions during the second half of the century means that Britain is home to overseas and first and second generation UK-born citizens with a variety of ethnic group and religious backgrounds" (see also Simpson 2012). However, the majority of Jews in Britain are now third- to fifth-generation UK-born (Waterman and Kosmin 1987). With the current residential tendencies, the building of new synagogues proved to be costly, and the requirement of Jews can be satisfied by small, multi-purpose buildings for up to 300 members, to serve religious, educational and communal needs.

The Board of Deputies of British Jews (2012) describes the British Jewish community as "contains two very different subcommunities, even though the boundary between them is not always well defined. The Haredi part of the community exhibit visually and locationally distinct characteristics which ensure that it is growing at a significant rate and that it comprises an increasing proportion of the Jewish population of the UK" (see also Vulkan and Graham 2008; Graham 2013; Sapiro 2016). According to Staetsky and Boyd (2015), a very high fertility level leading to a rapid expansion of the Haredi community. The community is growing at 4.8% a year, now comprising 16% of the Jewish population. In 2015, the Haredi population of Britain was estimated to be about 40,000. This population is characterised by younger age profile, early marriage and an average birth rate of seven children per woman (the national average is 1.93). Most of their children are sent to independent educational institutes. The lifestyle of the Torah scholars was supported indirectly by the British governments found, and people got used to dedicating their life to Torah study while living on benefits. Individuals and households got used to living on benefits; they became dependent (Spicker 2002). This phenomenon increases the poverty of Haredi people and reinforces their solidarity with the community's values and needs (Friedman 1991). Some 73% of British Haredi lives in London as a distinct minority. About half are affiliated with the Litvish group, and nearly half of these belong to the middle class (Waterman and Kosmin 1988). There are four main concentrations of the Haredi in Britain today: Stamford Hill/South Tottenham and Golders Green/Hendon areas in London; the Salford area in north Manchester and the unique Gateshead community in north-east England (Olsover 1980). The literature on religious transnationalism explains that the Haredi community in the UK appears to have weakened its orthodox position through sociocultural integration into the general society (Kislev 2015). However, the apartment prices in the Haredi enclaves are significantly higher than similar apartments among the general population, indicating the importance of spatial isolation and the need of living among "friends" for the Haredi sector.

There is considerable heterogeneity among "mainstream" Jews (Abramson et al. 2011). The "mainstream" community, the secular and moderately religious Jews, consist the United Synagogues and Federation synagogues, the Sephardi (3%), the Masorti movement (3%) and the Reform (19%) and Liberal (8%) synagogues. The United Synagogues and Federation synagogues cater the religious and traditional groups within the community. Within the groups, there is a common desire for community-oriented religious and social services, and these groups are concentrated in loosely defined spatial segregation in the north-west suburbs of London and the suburbs of other provincial towns. The few Masorti, Reform and Liberal synagogues also show a marked concentration in the northern suburbs. Due to the looser spatial and residential ties of this subgroup, they can manage with a fewer total number of synagogues for a membership drawn from a much wider area. The most recent development of small new communities in the towns of the Thames Valley indicates the potential of the Reform and Liberal Movements to draw its members from a wider hinterland. Sapiro (2016) examines the "spatial variation in the characteristics of "mainstream" Jews, show a clear division between inner urban Jews and suburban/commuter-belt/coastal Jews, with distinct classes within each of these two broader categories" (see also Becher et al. 2002; Graham et al. 2014). In between the Haredi and the "mainstream", apparently overlaps with the Religious Zionism, Federation and with Haredi Judaism at its respective boundaries, the movement of modern orthodox attempts to synthesise Jewish values and the observance of Jewish law with the modern world.

Friedman (1991) explains that from the point of view of an orthodox Jew, the practice of less orthodox Judaism is a mark of Jewish decline; from a sociological angle, however, attempts to "modernise" Judaism and make its practice easier for Jews caught up in a Gentile environment can be taken as evidence that the religion is surviving by adapting. In this regards, the structural entity and the demography of Anglo-Jewry are vague because there is no Jewish culture involving the greater part of all the individuals commonly accepted as being Jews. Freedman (1962) explains that Anglo-Jewry is a category of people in which every individual shares some Jewish characteristics with many other individuals but which is not uniform in its Jewish properties: "Jewish cultural habits there are in plenty, but they are not integrated or widespread enough to constitute a specific way of life peculiar to all or most Jews. Haredi Jews, for example, practice very different family formation norms to Jews generally, particularly with early marriage and very high birth rates. These result in quite different household structures for the rest of the Jewish population". Also, there is no general Jewish language. Hebrew as a modern tongue is sparsely known and used, mainly within the Israeli newcomers. Apart from the Hasidim, Yiddish has largely disappeared as an unspoken language even among the immediate descendants of the immigrants from Eastern Europe. Ladino speakers are numerically negligible. Unlike the case of Americans Jews, it is impossible to say that Anglo-Jews have developed their brand of English. Freedman (1962) explains that in Britain, without practising Judaism, a Jew can be Jewish by being a secular nationalist. However, the purely secular forms of Zionism are not very conspicuous, especially in recent decades, "with the rise in criticism of Israel's domestic and foreign policies and questioning the feasibility of Aliya"—the immigration of Jews from the diaspora to the Land of Israel—and support for Israel from both religious and not very religious Jews there is very common.

2.4 The Development of the Jewish North-Western Route

Data obtained from 1695 census and after that identified the precise location of the Jews population and indicated that over the years the development of the community had followed a specific pattern. Following Stowe's survey of London, Fig. 2.1 shows

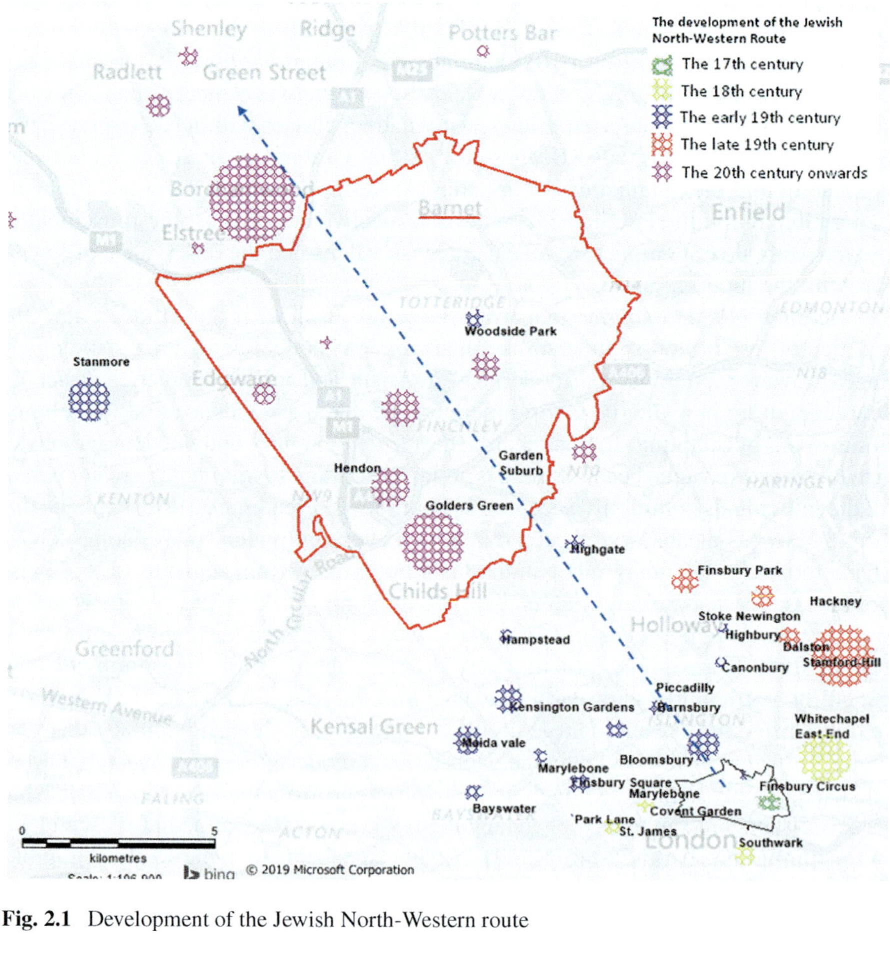

Fig. 2.1 Development of the Jewish North-Western route

that in 1695 there was a tendency to concentrate (in light green): 70% of the 750 Jews who can be identified lived in five contiguous parishes of St. James' Duke's place, St Katherine Cree, St. Andrew Undershaft, St. Katherine Coleman and St Helen's, within the old city of London (Newman 1976). During the eighteen century, the Jewish population of the metropolis increased up to 20k (Colquhoun, police of the Metropolis, 1800, p. 120), and the spatial composition of the Jewish community had changed. Figure 2.1 shows (in yellow) that in addition to the main Jewish settlement in the East End of London, two small Jewish settlements had been established in the West End around Covent Garden and St. James and Southwark, populated mainly by shopkeepers, craftsmen and few wealthier merchants.

During the eighteen century, the move from the City and East End began following few affluent families such as the Rothschild who moved from New Court to Piccadilly and Montefiore to Park Lane in about 1825 (in blue). Also, while retaining their primary homes in or near the city of London, the wealthiest Jews were also acquired country homes outside the city. Although the Jewish participation in the movements of residences from the City before to the West End had been minimal, the upper middle class followed this tendency supported by the new development of the top and the underground railway. The first new areas towards which the Jewish middle class moved as early as the 1830s were Finsbury Square, Finsbury Circus, Bloomsbury, Marylebone and Bayswater area, West of Marble Arch. In the 1830s and 1840s, Bloomsbury became the Jewish area of residential quality for the upper and the upper middle class and also for some of the less affluent. Throughout the nineteenth century, Bloomsbury was known for its considerable Jewish settlement, but by 1860 was being overtaken by the Bayswater. From about 1840 middle-class Jews from the city began to settle in some numbers in Islington, North London. By 1861, Islington had nearly 1000 Jews, about 5% of London's Jewish population. The Sephardim settled particularly in the north-west, in Stanmore, Totteridge, Watford, Hampstead and Highgate. The Bayswater Synagogue was founded in 1863, the same year with the opening of the first underground line from Paddington to the city. During the following years, Jews also moved into Canonbury and Highbury, to the north and east of the original settlement around Barnsbury where the north London synagogue was open in 1868. In 1870, a tendency for a spatial expansion of London Jewry out of the small areas of settlement into the suburbs was followed by a further movement towards the west, along with the north side of the Park and Kensington Gardens. A few years later, a more popular residential area for a less affluent middle class was opening in Maida Vale. But a few decades later, from 1890 onwards, Jews move out of Maida Vale into the newer residential areas of the north-west.

From 1860, the second central axis was northwards along the road first laid out by the Romans (known subsequently as Ermine Street) and through Barnsbury, Canonbury, Highbury, Dalston, Stoke Newington and Finsbury Park. The foundation of St. John's Wood's synagogue in 1875 was the beginning of Jewish settlement in the north-west. By 1880, the upper middle class had already moved out of the traditional area of Whitechapel and other parts of the East End, while most of the middle-class Jews were in a moving process. In addition to Barnsbury, Canonbury and Highbury, other further east neighbourhoods, such as Dalston and Hackney, had started to attract Jewish residents.

1880 marks the beginning of the mass influx of Eastern European Jews immigrated into Britain in the last decades of the nineteenth century, which has dramatically transformed Anglo-Jewry. The establishment of the Jewish enclave of Whitechapel in the East End (that held about 90% of metropolitan Jewry) and some other provincial centres as replicas of the continental compact settlements maintained the community characteristics even in the face of the crisis in the late 1880s (Flint Ashery 2018a, b). But when the Eastern European Jews newcomers start exploiting the business opportunities open to them and making good use of the public education freely available, many Eastern European Jews rose to middle-class status and moved to the new Jewish areas, and the English "ghettoes" did not continue this trend. Inside London, and after the First World War, Jews flowed along this northern route which marked various stages in the process of social as well as physical mobility. Jewry in Britain has as a whole a decidedly middle-class complexion, and Jews are proportionally higher than non-Jews working on their account. For many households working on their account in their small- or medium-scale business, the relatively flexible nature of their work helps in this spatial process. As the Jewish households grew more prosperous, they moved further north-west. The relocation of the residential housing of the Jewish community to the suburbs has led to a parallel relocation of the various community institutions. This process is in line with Ward (1982) explanation that "one of the most defensive contributions to structural pluralism was the maintenance of ethnic institutions". Groups which move to the suburbs develop a strong association with the relocation of ethnic institutions, around which community attachment is maintained.

With the influx of the Haredi population in the UK, this one index of the essential and primary character of the Jewish life in Britain—their need to live among people of their group—has changed (Freedman 1962). Haredi is by far the fastest-growing component of contemporary British Jewry and Jewry worldwide (Belcove-Shalin 1995; El-Or 1994). Vulkan and Graham (2008) explain that "the expansion of the Haredi sector is also influencing the population level of London Jewry but remains confined to specific enclaves". Based on his work of Chicago, Wirth (1927/1998) recounts the story of Medieval, Renaissance and eighteenth-century Jewish quarters leading up to the nineteenth-century ghetto and shows how Jews have a natural tendency to voluntary concentrate in particular areas of the city. In this regards, the Jewish Heritage explains that the Jewish community is "concentrated in a specific urban and suburban areas, in north-west London and north-west Manchester. This pattern is resulting in the disappearance of smaller communities around the country, in both small towns and large cities, including Liverpool, Leeds, Birmingham and Glasgow".

Nationally, Britain's Jewish population today comprise just 0.4% of the national population, concentrated mainly in the large urban centres, in a few small localised areas and neighbourhoods in London, Manchester, Leeds, Glasgow, Liverpool, and Birmingham. This residential tendency is mainly due to the support and infrastructures that are available there. However, smaller communities have mostly developed

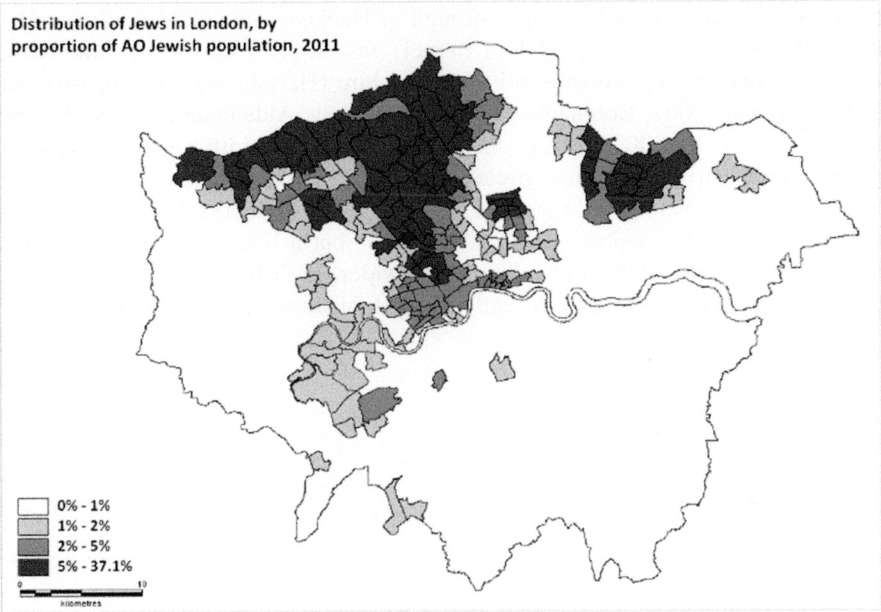

Fig. 2.2 Distribution of the Jewish population in London

in areas of industry and survive due to one particular circumstance (Brook et al. 1989). According to the 2011 census, at the local level, on the other hand, the proportion of the Jewish population rises to as much as 40% in some places. Six of the largest Jewish neighbourhoods in 2011 contain 5000 or more Jews. According to Staetsky and Boyd (2015) "four wards—Kersal and Sedgley (in Greater Manchester), and Springfield and New River (in Hackney)—the Jewish population is predominantly Haredi".

56.9% of the Jewish population lives in London (Fig. 2.2). Inside the capital, the London borough of Barnet has the most substantial proportion, with 20.5% of the Jewish community of England and Wales living there. Within the twenty-one neighbourhoods (or wards) of Barnet, Golders Green is the most Jewishly populous with 7661 Jews. Other neighbourhoods are Hendon, Hampstead Garden Suburb and Finchley, all of which have sizeable Jewish populations. Fox (2016) describes how Golden Green, Hendon and Edgware have marked terminal points of the migratory route from the East End: "although Golders Green was ranked in the third largest ward in 2001, it was in the top spot in 2011. Between 2001 and 2011, Golders Green's Jewish population increased by over a third (34.6%), and today Golders Green is the largest Jewish neighbourhood in the country (a position formally held by Garden Suburb ward)". The census data also provide an indication of Jewish population change over time at the local level.

In the case of the Jewish community in London, Jewish Heritage (2015) explains how the community maintained its segregation through the process of simultaneously moving their institutions with them: "In addition to the large Jewish communities of

Stamford Hill and Clapton (in the Borough of Hackney) and Golders Green, Hendon and Edgware (in the Borough of Barnet), fastest-growing Jewish communities in the country are located in South Hertfordshire (Hertsmere)". The census indicates that since 2001, Borehamwood Brookmeadow, Aldenham East and Shenley have grown by 40–50%. Most of these areas populated with a preponderance of young families affiliated to mainstream synagogue organisations, who were seeking a Jewish environment of a higher social standing. It is important to note that this route does not show a procession of "ghettoes" but a line of Jewish areas which are evaluated partly by their standing in the wider English world and the existence of Erov wire (Fig. 2.3). Only recently, the establishment of Woodside Park Eruv

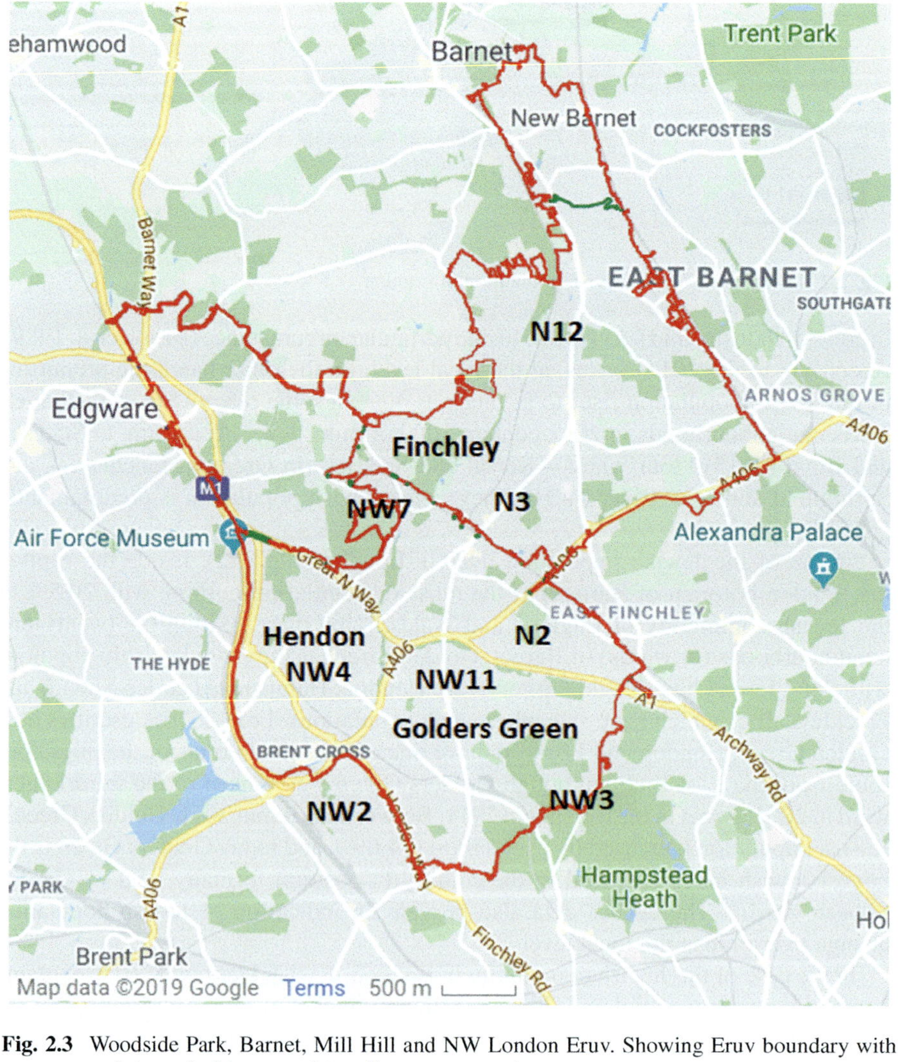

Fig. 2.3 Woodside Park, Barnet, Mill Hill and NW London Eruv. Showing Eruv boundary with crossover points marked in green. https://www.eruv.co.uk

(https://www.woodsideparksynagogue.org.uk/eruv), for example, attracted many new Jewish families to the area, west Finchley Chabad house was founded, and the prices of properties in the area increased compared to other nearby areas without Eruv. We can, therefore, assume that Freedman (1962) assumption that the Jewish like to live among Jews but not to the extent of cutting themselves off from the wider society and such a spontaneous route enables them to circulate freely and equally is still relevant for the mainstream Jewish population. In addition to these "London related new emergent communities", the ward of Radcliffe West in the west of Bury, Greater Manchester, and Lobley Hill and Bensham in Gateshead have grown by almost 150% since 2001 (Jewish Heritage UK 2015). Graham (2013) explains that in particular, "the Haredi ward of Sedgley in Bury grew by almost 42% from 2001 to 2011 while the Haredi ward of Seven Sisters in Haringey doubled its size, increasing by an astonishing 7.3% per year on average since 2001". Most of these areas populated with a preponderance of young families affiliated to Haredi synagogue organisations, who were seeking to rely on their home community (Holman and Holman 2002; Valins 2003a, b).

2.5 Haredi Jews in the UK and Their Relations with the General Society

During the years, the demography, the sociology and the relationships between the of Anglo-Jewry and the general society have been included in more extensive studies of British Jewry (Brook et al. 1989; Cooper and Morrison 1991) shows that the Jewish community had come a long way from the poverty of the early nineteenth century. The 1830s was the age of emancipation, and not only the wealthiest but also the middle-class Jews felt the desire to participate in trends affecting the wider community. When the communal leadership was engaged in the struggle for political emancipation, the Jewish community has assimilated thoroughly within the host community (Eisenberg 1996; Vincent and Warf 2002). An early example of this can be seen in the Queen's participation in a reception in the centenary of the United Synagogue (1870-) was commemorating, and the service for Thanksgiving that took place in St. John's Wood Synagogue. The Jews in the UK had been gradually accepted both politically and socially into almost every sphere of activity and had followed the patterns of the wider host community. With a growing desire to feel themselves to be Englishmen, many middle-class Jews followed their non-Jewish neighbours of similar social and economic status and moved to a more spacious square and terraces of central London.

To an outside observer, the Jewishness of Jews in Britain must often appear a rather queer amalgam of Judaism and some Jewish cultural oddities. In a meaningful sense, Anglo-Jewry is the product of British society. The growth of Jewish participation in business and industry, as well as professions such as medicine, the law and architecture, reflects the increase of a Jewish middle class. Parallel with a similar development in the outside world, in this process that has repercussions in the organisation of the community and the leadership of Anglo-Jewish society, the

Jewish absorbed the habits and costume of the wider non-Jewish society. The lay leadership of the Jewish community was an English institution that involved heavily in the Jewish and non-Jewish welfare and communal activities. Casale Mashiah and Boyd (2017) argue that "there is some sort of connection between the practice of the various forms of the religion and general class position, while the shifts in ritual and theology made in the name of reform are aspects of the process of anglicisation". As it has developed within the framework of this society, Anglo-Jewry has come to have little corporate existence, and its lack of cultural vigour may well be the price it pays for its freedom from the external pressure to constitute a political and legally defined body. Taking part and highly integrating within the general society, the most important contemporary Jewish anxiety is that of Jewish continuity. In particular, out-marriage (estimated to be over 50%) is a threat that demonstrates the role of the communities' institutions in increasing the intercommunal relations, thus to maintain the custom of marrying only within the limits of a Jewish community. The second most important anxiety is that of anti-Semitism. In 2014, the Department of Communities and Local Government described that the majority of anti-Semitic hate crime are focused in Metropolitan, Greater Manchester and Hertfordshire (DCLG 2015). In 2015, the CST found that 45% of British Jews feared they may have no future in Britain, and 77% of British Jews had witnessed anti-Semitism disguised as a political comment about Israel (CST 2016). Despite the indications, JPR found in 2017 that the levels of anti-Semitism in the UK were among the lowest in the world, with about 70% having a favourable opinion of Jews (Staetsky 2017).

While Haredi Jewry has been the source of considerable academic fascination in the USA and Israel, Haredi in the UK has received remarkably little scholarly attention from geographers. There have been a number of neighbourhood surveys of the local Jewish community, including such areas as Edgware (Krausz 1969a, b); Wembley (Cromer 1974); the inner city of Leeds (Grizzard and Raisman 1980); Dublin (Waterman 1981; 1984); Sheffield (Kosmin et al. 1976); Inner London (Kosmin and Grizzard 1975); and Redbridge (Kosmin et al. 1979), and Golders Green (Fox 2016). Even though, as Valins (2000) argues about twenty years ago, "a quarter of all Jewish school-age children are now Haredi and that this will have significant repercussions on the future characteristics of British Jewry", academic studies explicitly looking at British Haredi have tended to concentrate on the politics of constructing Eruvin or else on strategic issues such as housing and welfare needs (Holman and Holman 2001, 2002). Competition over space and resources between the Haredi and the general population already takes place at the local level, and researches assume that with the growth of the Haredi population, competition among groups may increase. In an attempt to minimise conflicts between the Haredi community and other social groups, the central and local governments are sometimes involved in the construction and management of Haredi residential areas. Also, local governments often provide the necessary regulation for maintaining the Haredi public space; this includes setting Eruv columns and allocating buildings for religious services. However, the difficulty of enumerating the members of Haredi communities as we shall see in the next chapter, and identifying their unique needs, sometimes impairs the ability of the authorities to meet their needs.

References

Abramson S, Graham D, Boyd J (2011) Key trends in the British Jewish community: a review of data on poverty, the elderly and children

Alderman G (2011) Jews and electoral politics in the United Kingdom: a contemporary note. Jew J Sociol 52(1)

Alfasi N, Flint Ashery S, Benenson I (2013) Between the individual and the community: residential patterns of the Haredi population in Jerusalem. Int J Urban Reg Res 37(6):2152–2176

Aran G (2013) On religiosity and super-religiosity (II): the case of Jewish ultra-orthodoxy. Numen 60(4):371–419

Aspinall P (2000) Should a question on 'religion' be asked in the 2001 British census? A public policy case in favour. Soc Policy Admin 34(5):584–600

Becher H, Waterman S, Kosmin B, Thomson K (2002) A portrait of Jews in London and the South-East

Bechhofer YG (1998) The contemporary Eruv: Eruvin in metropolitan areas

Belcove-Shalin JS (1995) Home in exile: Hasidim in the new world. In: New world Hasidim: ethnographic studies of Hasidic Jews in America, pp 205–236

Ben-Sasson HH (ed) (2008) A history of the Jewish people. Recording for the Blind & Dyslexic

Berman E (2000) Sect, subsidy, and sacrifice: an economist's view of ultra-orthodox Jews. Q J Econ 115(3):905–953

Blumen O (2007) The gendered display of work: the midday scene in an ultra-orthodox street in Israel. Nashim J Jewish Women's Stud Gender Issues (13):123–154

Brook JS, Nomura C, Cohen P (1989) A network of influences on adolescent drug involvement: neighborhood, school, peer, and family. Genet Soc Gen Psychol Monogr

Casale Mashiah D, Boyd J (2017) Synagogue membership in the United Kingdom in 2016

Chakak Y (2004) Vocational training for ultra-orthodox men. The Floersheimer Institute for Policy Studies, Jerusalem (in Hebrew)

Cooper H, Morrison P (1991) A sense of belonging: dilemmas of British Jewish identity. Weidenfeld and Nicholson

Cromer G (1974) Intermarriage and communal survival in a London suburb. Jew J Sociol 16(2):155–169

CST (2016) Antisemitism barometer. Campaign Against Antisemitism. Retrieved from 2016-10-17

Dahan M (1999) The ultra-orthodox Jewish and municipal authority: part 2—income distribution in Jerusalem. The Jerusalem Institute for Israel Studies, Jerusalem

DCLG (2015) GOV.UK. Department for communities and local government Britain. www.gov.uk/government/uploads/system/uploads/attachment_data/file/390904/GovernmentAction_on_Antisemitism_final_24_Dec.pdf. Retrieved from 17 Jan 2015

Eisenberg R (1996) Boychiks in the hood: travels in the Hasidic underground. HarperSanFrancisco

El-Or T (1994) Educated and ignorant: ultraorthodox Jewish women and their world. Lien Rienher, Boulder

Etkes E, Tikochinski S (2004) Memoirs of the Lithuanian Yeshiva, Immanuel (eds.). Jerusalem, the Zalman Shazar Center for Jewish History

Finke R (1997) The consequences of religious competition: supply-side explanations for religious change. In: Young LA (ed) Rational choice theory and religion: summary and assessment. Routledge, NewYork, NY

Finke R (2015) Religion and struggle in the European Union: confessional culture and the limits of integration. By Nelsen Brent F. and Guth James L. Georgetown University Press, Washington, DC

Flint Ashery S (2018a) Micro-residential dynamics. A case study of Whitechapel, London. Springer. ISBN 978-3-030-00601-3

Flint Ashery S (2018b) Schelling-type micro-segregation in a Hasidic enclave of stamford-hill. Hous Stud. https://doi.org/10.1080/02673037.2017.1414160

Flint S, Benenson I, Alfasi N (2012) Between friends and strangers: micro-segregation in a Haredi neighborhood in Jerusalem. City Community 11(2):171–197
Fox P (2016) The Jewish community of golders green: a social history. History Press, Stroud. ISBN 978-0-7509-6587-3
Freedman M (1962) The Jewish population of great Britain. Jew J Sociol 4(1):92–100
Friedman M (1991) Ha'hevra Ha'haredit—Mekorot, Megamot, Tahalichim [The Haredi ultra-orthodox society: sources, trends and processes]. Research series no. 41. The Jerusalem Institute for Israel Studies, Jerusalem
Godley A (1997) The development of the clothing industry: technology and fashion
Gonen A (2006) Bein Limud Torah U'parnasa: Hevrat Lomdim Umitparnesim Be'London [Between Torah learning and wage earning: the London experience and lessons for Israel]. The Floersheimer Institute for Policy Studies, Jerusalem
Graham D (2013) 2011 census results (E&W): a tale of two Jewish populations. Institute for Jewish Policy Research, London
Graham D, Staetsky LD, Boyd J (2014) Jews in the United Kingdom in 2013: preliminary findings from the National Jewish Community Survey. Institute for Jewish Policy Research, London
Green A (2001) Typologies of leadership and the Hasidic Zaddiq. In: Assaf D (ed) Zaddik and devotees: historical and social aspects of Hasidism. The Zalman Shazar Center for Jewish History, Jerusalem, pp 422–444 (in Hebrew)
Grizzard N, Raisman P (1980) Inner city Jews in leeds. Jew J Sociol London 22(1):21–33
Heritage UK. Author(s): Jewish Heritage UK. Date: 2015
Holman C, Holman N (2001) Orthodox Jewish housing need in Stamford Hill
Holman C, Holman N (2002) Torah, worship and acts of loving kindness: baseline indicators for the Charedi community in Stamford Hill. Interlink Foundation
Interlink (2010) The spatial needs of the Charedi community in Stamford Hill
Kislev E (2015) The transnational effect of multicultural policies on migrants' identification: the case of the Israeli diaspora in the USA. Glob Netw 15(1):118–139
Kosmin BA, Grizzard N (1975) Jews in an inner London borough. Board of Deputies, London
Kosmin BA, Bauer M, Grizzard N (1976) Steel city Jews: a study of ethnicity and social mobility in the Jewish population of the city of Sheffield, South Yorkshire
Kosmin BA, Levy C, Wigodsky P (1979) The social demography of Redbridge Jewry. Research Unit, Board of Deputies of British Jews
Krausz E (1969a) Sociology in Britain: a survey of research. Batsford, London
Krausz E (1969b) Edgware survey-occupation and social class. Jew J Sociol 11(1):75–95
Lupo J (2004) Can Shas restore past glory? The Floersheimer Institute for Policy Studies, Jerusalem (in Hebrew)
Newman A (1976) The United Synagogue, 1870–1970. Routledge and Kegan Paul, London [etc.]
Olsover L (1980) The Jewish communities of North East England 1755–1980. Ashley Marks Publications Co, Newcastle Upon Tyne
Phillips D (2006) Parallel lives? Challenging discourses of British Muslim self-segregation. Environ Plan D Soc Space 24(1):25–40
Piekarz M (1999) Ha'Hanhaga Ha'hasidit [The Hasidic leadership authority and faith in Zadicim]. Bialik Institute Publishing House, Jerusalem
Sapiro P (2016) Beyond the Haredi/mainstream divide: applying geodemographic analysis to a small nationwide sub-population. Comput Environ Urban Syst 56:36–47
Shilhav Y (1993) The emergence of ultra-orthodox neighborhoods in Israeli urban centers. Local Communities Israeli Polity, 157–187
Shilhav Y, Friedman M (1985) Growth and segregation: the Haredi community of Jerusalem. The Jerusalem Institute for Israel Studies, Jerusalem
Simpson L (2012) More segregation or more mixing? The dynamics of diversity: evidence from the 2011 census. Centre on Dynamics of Ethnicity (CoDE), Manchester
Sivan E, Caplan K (2003) Haredim Israelim: Hishtalvut lelo Tmiah? [Israeli Haredim: integration without assimilation?]. Hakibbutz Hameuhad, Tel Aviv

Spicker P (2002) Poverty and the welfare state: dispelling the myths. Catalyst, London

Staetsky LD (2017) Antisemitism in contemporary Great Britain. Institute for Jewish Policy Research (JPR)

Staetsky D, Boyd J (2015) Haredi rising: what the demography of British Jews tells us about the future of the community. Institute for Jewish Policy Research (JPR)

Tausch N, Hewstone M, Schmid K, Hughes J, Cairns E (2011) Extended contact effects as a function of closeness of relationship with ingroup contacts. Group Process Intergroup Relat 14(2):239–254

Valins O (2000) Institutionalised religion: sacred texts and Jewish spatial practice. Geoforum 31(4):575–586

Valins O (2003a) Defending identities or segregating communities? Faith-based schooling and the UK Jewish community. Geoforum 34(2):235–247

Valins O (2003b) Stubborn identities and the construction of socio-spatial boundaries: ultra-orthodox Jews living in contemporary Britain. Trans Inst Br Geogr 28(2):158–175. https://doi.org/10.1111/1475-5661.00085

Vaughan L, Penn A (2006) Jewish immigrant settlement patterns in Manchester and Leeds 1881. Urban Stud 43(3):653–671

Vincent P, Warf B (2002) Eruvim: talmudic places in a postmodern world. Trans Inst Br Geogr 27(1):30–51

Vulkan D, Graham D (2008) Population trends among Britain's Haredi Jews: board of deputies of British Jews

Ward D (1982) The ethnic ghetto in the United States: past and present. Transactions of the Institute of British Geographers, pp 257–275

Waterman S (1981) Changing residential patterns of the Dublin Jewish community. Irish Geogr 14(1):41–50

Waterman S (1984) Partition—a problem in political geography. In: Political geography: recent advances and future directions, pp 98–116

Waterman S, Kosmin BA (1987) Residential change in a middle-class suburban ethnic population: a comment. Trans Inst Br Geogr 12(1):107–112

Waterman S, Kosmin BA (1988) Residential patterns and processes: a study of Jews in three London boroughs. Trans Inst Br Geogr 13(1):79–95

Wirth L (1927/1998) The ghetto. Am J Sociol 33(1):57–71

Yiftachel O, Roded B, Kedar A (2016) Between rights and denials: Bedouin indigeneity in the Negev/Naqab. Environ Plan A Economy Space 48(11):2129–2161

Zelkin M (2006) Social status and authority in 19th-century Lithuanian Jewish communities. In: Verbickiene J (ed) Central and East European Jews at the crossroads of tradition and modernity. The Center for Studies of East European Jews, Vilnius

Chapter 3
Measurement of the Jewish Population in the UK

Abstract The Haredi are reluctant to disclose information that enables authorities and researchers to estimate the size of the population, as we are not allowed to count the people or mention names. These factors, together with limited fluency in English among the Hasidim, suggest that underenumeration of the Haredi population is inevitable. During the years, two data sources—synagogue membership records and the "North London Shomrei Shabbos [lit. Sabbath Observers] Directory"—had been the most critical data that exist about Jews in the UK. Other invaluable sources of data are the annual Jewish Year Book lists the number, and location, of all Jewish institutions throughout the country as well as containing a mass of information concerning the community as a whole; the weekly Jewish Chronicle is another invaluable source of community information dating back some 150 years in addition to The Yeshiva world, Jewish telegraphic agency JTA and the Jewish small communities network; and the notices of Shalom Zachar celebrations, which take place in Ashkenazic Jewish circles following the birth of a male child. Each data source has its methods according to its initiation. The debate over changes to the 2001 census has led to the inclusion of a question on religion in the England and Wales 2011 census. As a person's religion is considered personal information, answering this question is voluntary due to the Human Rights Act 1998. This study, however, is mostly built on first-hand data. Over twelve years of research, since the beginning of my doctoral dissertation, I have been intensely involved in the Haredi community in Israel and London. Being deeply familiar with the lifestyle and ritual orientation of each group, and my strong ties with the members of the various communities have opened many doors to this unique world.

Keywords Data · Synagogue membership records · North London Shomrei Shabbos Directory · Shalom Zachar · 2011 census

S. Flint Ashery, *Spatial Behavior in Haredi Jewish Communities in Great Britain*,
The Urban Book Series, https://doi.org/10.1007/978-3-030-25858-0_3

3.1 The Difficulty of Enumerating Haredi Jews

The size of the UK Jewish population has always been a source of uncertainty for demographers, and the main difficulty encountered in studying the Haredi society entailed locating information. Orthodox Jewish law recognises the biblical injunction in Exodus Chapter 30 (12–13), against the physical counting of Jews:

> 12 "When thou takest the sum of the children of Israel, according to their number, then shall they give every man a ransom for his soul unto the LORD, when thou numberest them; that there be no plague among them, when thou numberest them."
>
> 13 "This they shall give, every one that passeth among them that are numbered, half a shekel after the shekel of the sanctuary–the shekel is twenty gerahs–half a shekel for an offering to the LORD."

The Haredi, according to Graham and Waterman (2005), are reluctant to disclose information that enables authorities and researchers to estimate the size of the population, as we are not allowed to count the people or mention names. Many Haredi people are more likely to be influenced by the biblical prohibition on counting Jews and are intolerant of being counted. In addition, they "generally adhere to the concept of da'at Torah which refers to the totality of rabbinic knowledge and interpretation a principle of significance if a rabbi decrees his congregants not to answer the voluntary census question on religion." In this regards, Alderman (2008) explains that most of the Haredi communities, both Litvish and the Hasidim, do not recognise the authority of the British Board of Deputies of Jews, and they ignore or refuse any requests from the BoD's leaders to be counted. These factors, together with limited fluency in English among the Hasidim (many do not speak English as a first language even if they were born in the UK, see Holman and Holman (2001, 2002)), suggest that underenumeration of the Haredi population is inevitable.

Attempts to gather statistical, demographic or any other information are impeded because of the Haredi sector—and all the more so its groups—is not an independent category and there are no separate statistical or demographic records for it, so that quantitative data referring to the population are likely to be based on estimate, assessment or survey (Flint Ashery et al. 2012; Flint Ashery 2014). Besides, since 2001 anyone in Britain is entitled to remove himself from the databases. Hence, for many research, any up-to-date quantitative data on this community is estimated by Yeshiva data or publication of informal community notices.

Another problem in enumerating Haredi Jews relates to suspicion of how the data might be used. Weller and Andrews (1998) explain that "Historically, many Jews have been suspicious of government-sponsored questionnaires requiring them to identify their religion. This suspicious can be related directly to a history of persecution of which many older members of the community still have personal memories." Ethical concerns focus upon the delicate nature of dealing with sensitive issues surrounding a minority faith or ethnic group. These concerns are for two reasons, firstly, to protect the community and secondly, to allow research to continue in the future (Clifford and Valentine 2003:39).

3.2 Unofficial and Official Data Sources About Jews in the UK

During the years, two data sources—synagogue membership records and the "North London Shomrei Shabbos [lit. Sabbath Observers] Directory"—had been the most critical data that exist about Jews in the UK. Other invaluable sources of data are the annual Jewish Year Book lists the number, and location, of all Jewish institutions throughout the country as well as containing a mass of information concerning the community as a whole; the weekly Jewish Chronicle is another invaluable source of community information dating back some 150 years in addition to The Yeshiva world, Jewish telegraphic agency JTA and the Jewish small communities network; and the notices of Shalom Zachar celebrations, which take place in Ashkenazic Jewish circles following the birth of a male child. Each data sources has its methods according to its initiation.

The synagogue membership records, in this regards, reflect the relatively high propensity of Jewish males, mainly of the non-Haredi groups, to be members of a synagogue. The synagogue membership is distributed over a range of congregations which span the ritual and theological gap between extreme orthodoxy and Liberal Judaism. However, Casale Mashiah and Boyd (2017) explain that as synagogue membership statistics data are "gathered directly from synagogues, there is no way to ascertain whether any members accounted for in one synagogue's counts are also considered for in another's." This tendency is more "problematic" for the Hasidic community due to the common practice not to count individuals directly. The gap between the proportion of Jewish adult men who make use of the synagogues and the actual number of members of synagogues indicates that about third of all Jewish adult men are members of synagogues. The data suggest that "in 2016, 79,597 Jewish households across the UK held mainstream synagogue membership. About 64% of all synagogues in the UK is situated in Greater London and the adjacent areas of South Hertfordshire and South-West Essex (10%)."

JPR (2015) argues that "in the UK synagogues have long been the primary and most common means by which Jews affiliate to the Jewish community." The "North London Shomrei Shabbos Directory", provided a complimentary picture with the address details of most Haredi Jewish households in Stamford Hill vicinity, (Gonen 2006) and has been, therefore, the main source for estimating the Haredi community's demographic characteristics. According to Vulkan (2012), "the 2012's Shomer Shabbos directory for London recorded half as many households in North West London as in Stamford Hill which, by proportion, would suggest the birth of 237 Haredi Jewish boys in 2012, leading to a total of 962. Assuming that Haredi families in North West London tend to have fewer than those in Stamford Hill, a reasonable estimate would be of 800–900 boys in total, or around 1600–1800 children in total."

Notices on Shalom Zachar celebrations are published before the Friday night after the boy is born, to invite the neighbourhoods' children to pray Shema next to the new baby. For Stamford Hill and other Yiddish-speaking Hasidic community, the notices of Shalom Zachar celebrations are published weekly in the Kol Mevaser

Yiddish broadcasting of the neighbourhood. The Litvish circles distribute this event directly in the synagogue (e.g. Gateshead Hebrew Congregation and Gateshead Kolel Synagogue). Reviewing Shalom Zachar notices, Vulkan (2012) explains that as the amount of male births recorded for the year 2012 "is comprised of 474 in Stamford Hill and 209 in Manchester. This would infer a total of over thousand births of both male and female in 2007, rising to over 1300". For this period, an annual increase of 4.6% (in average) indicates the growth of the Haredi population in the UK. With regard to the Litvish community, Casale Mashiah and Boyd (2017) find that according to the Gateshead Kehilla Directory (2013 estimated year of publication), Haredi membership figures have increased dramatically: "by 139%, rising from a recorded count of 4489 in 1990 to 10,712 in 2016, the equivalent of about 240 new households per annum. Once an allowance is made for the communities in North West London and Gateshead, a total figure of between 1600 and 1800 would not be unreasonable, and a conservative estimate would be that the Haredi community now accounts for 40% or more of all Jewish births."

The debate over changes to the 2001 census has led to the inclusion of a question on religion in the England and Wales 2011 census. As a person's religion is considered personal information, answering this question is voluntary due to the Human Rights Act 1998. However, and in spite of the voluntary nature of this question, this data allow "a cross-tabulation of the socio-economic, demographic and household structure characteristics of Anglo-Jewry in 2011 to be carried out. The 2011 census recorded 211 Jewish children under the age of 5 in Gateshead, suggesting around 42 boys being born in a year, bringing the total to 725" (Sapiro 2016). The data indicate that overall, the Jewish population, the oldest non-Christian faith minority in Britain, has changed dramatically. According to the Jewish Heritage report (2015), the population "has almost halved in size from a reputed peak of 450,000 in the 1950s. Initial estimates put the total number of Jews in Britain at 263,000, down from 267,000 recorded in the 2001 census, and up to 2,712,952 according to more recent estimation" (see also Staetsky and Boyd 2015). The current Jewish community is less than half of one per cent of the total population of the UK. This overall decline was attributed mainly to a drop in the birth-rate in the mainstream of Anglo-Jewry—that follows the common pattern of fertility in middle-class Britain in general—resulting in a rising age profile and an excess of deaths over births. Other significant factors are out-marriage and emigration, mainly to Israel. According to JPR (Staetsky and Boyd 2015) "whilst the Jewish population in England and Wales grew by 2% between 2001 and 2011, the number of households in England and Wales in which at least one Jewish person lives declined by 4.2% between 2001 and 2011. Assuming the same rate of decline across the UK as a whole for the period 2011 and 2016, generates an estimated total number of Jewish households for the UK in 2016 of 141,503."

This study, however, is mostly built on first-hand data. Over 12 years of research, since the beginning of my doctoral dissertation, I have been intensely involved in the Haredi community in Israel and London, even if I am not a member of the community myself. Being deeply familiar with the lifestyle and ritual orientation of each group, and my strong ties with the members of the various communities have opened many doors to this unique and impressive world. Despite the desire to expose the uniqueness of the community to the world, to be recognised and to reduce the level of alienation, many Haredi fear that they will waste Torah learning time and invest in collaborations that could result in manipulation or misuse of data against their community. In this context, forming long-term networks with people in the community through already established connections from parallel communities who are familiar with me and with my work, and maybe, being a woman with an Israeli accent, reduced the level of concern and raised the motivation for cooperation. Based on primary and secondary data, the next four chapters examine the implementation of behavioural/normative delineations, and their application in the urban realm of the Litvish communities in Golders Green and Gateshead, and the Hasidic communities of Stamford Hill and Canvey Island, expose internal trends that allowed for comparison between the two groups in terms of lifestyles, norms and residential preferences. Note that all census data and mapping used in this work are "Crown Copyright and reproduced or adapted from data from ONS (the Office for National Statistics) licensed under the Open Government Licence v.2.0, and all census tables referred to can be accessed via http://www.nomisweb.co.uk/census/2011/datafinder".

References

Alderman G (2008) Controversy and crisis: studies in the history of the Jews in modern Britain. Academic Studies Press

Casale Mashiah D, Boyd J (2017) Synagogue membership in the United Kingdom in 2016

Clifford NJ, Valentine G (2003) Key methods in geography. Sage, London

Flint Ashery S (2014) Between the individual and the community: residential behaviour of the Haredi population in Jerusalem, Haredi Society 1.8, pp 66–99, The Jerusalem Institute for Israel Studies (Hebrew)

Flint S, Benenson I, Alfasi N (2012) Between friends and strangers: micro-segregation in a Haredi neighborhood in Jerusalem. City Community 11(2):171–197

Gonen A (2006) Bein Limud Torah U'parnasa: Hevrat Lomdim Umitparnesim Be'London [Between Torah learning and wage earning: the London experience and lessons for Israel]. The Floersheimer Institute for Policy Studies, Jerusalem (in Hebrew)

Graham D, Waterman S (2005) Underenumeration of the Jewish population in the UK 2001 census. Population, Space Place 11(2):89–102

Holman C, Holman N (2001) Orthodox Jewish housing need in Stamford Hill

Holman C, Holman N (2002) Torah, worship and acts of loving kindness: baseline indicators for the Charedi community in Stamford Hill. Interlink Foundation

Sapiro P (2016) Beyond the Haredi/mainstream divide: applying geodemographic analysis to a small nationwide sub-population. Comput Environ Urban Syst 56:36–47

Staetsky D, Boyd J (2015) Haredi rising: what the demography of British Jews tells us about the future of the community. Institute for Jewish Policy Research (JPR)

Vulkan D (2012) Britain's Jewish community statistics 2010. Board of Deputies of British Jews, London, pp 1–17

Weller P, Andrews A (1998) Counting religion: religion, statistics and the 2001 census. World Faiths Encounter 21:23–34

Chapter 4
The Litvish World

Abstract "Litvish" is the name given to Jews identified with the religious tradition of Rabbi Elijah of Vilnius, and his disciples, who founded Yeshivas characterised by an ideology of religious conservatism. The Litvish community holds religious study as a central cultural value. Aside from identification with the group, communal identity is also channelled to exposing the younger generation, from an early age to the community's values and educating him within as homogeneous an environment as possible, as a means to continuity and strengthening of the community. Thus, the Litvish identity—along with the distinction between those "like us" and "others"—becomes a central component of the individual's consciousness, influencing his choices and preferences. Of these, the most significant is the sense of mission—preservation of Jewish survival—that community members feel by their religious status. This feeling bestows tremendous importance on educating the younger generation following the community's values and enhances the community's resilience. Common interests and communal identity create internal Litvish mechanisms that educate the younger generation and affect appropriate socialisation. Nevertheless, as we will see in the next chapters, life in modern urbanites, where most of Britain's Haredi population is concentrated, presents socio-economic and religious challenges.

Keywords Litvish · Rabbi Elijah of Vilnius · Da'at Torah · Religious elite · Yeshiva · Socialization process

"Litvish" is the name given to Jews identified with the religious tradition of Rabbi Elijah of Vilnius, and his disciples, who founded Yeshivas characterised by an ideology of religious conservatism. The imposition of 'da'at Torah'—a Litvish—Haredi doctrine awarding top-tier religious scholars the authority to rule on matters both private and public, even outside of Litvish circles—established the Litvish status as the Haredi "religious elite" that determines the norms of Haredi society (Etkes and Tikochinski 2004; Stampfer 2012). Although the consensus around the Litvish ideal has been weakened during the last decade, the positive attitude among other Haredi circles towards individuals and bodies affiliated with this group is expressed in such phenomena as the aspiration of youth from other groups to be accepted to Litvish Yeshivas; matchmaking inquiries; and a preference for housing in close proximity to the Litvish community (Lupo 2004). Despite external pressure, common interests

S. Flint Ashery, *Spatial Behavior in Haredi Jewish Communities in Great Britain*, The Urban Book Series, https://doi.org/10.1007/978-3-030-25858-0_4

and communal identity create internal Litvish mechanisms that educate the younger generation and affect appropriate socialisation. Generally, Litvish institutions avoid accepting members of other Haredi groups; marriages usually involve two families from within the wider Litvish community, and there is a preference for housing within a Haredi enclave that facilitates segregation from non-Litvish influences. These preferences are internalised from early childhood, with the explicit affiliation of the individual with his community.

Aside from identification with the group, communal identity is also channelled to exposing the younger generation, from an early age to the community's values and educating him within as homogeneous an environment as possible, as a means to continuity and strengthening of the community (Bilu 2003). Thus, the Litvish identity—along with the distinction between those "like us" and "others"—becomes a central component of the individual's consciousness, influencing his choices and preferences. Of these, the most significant is the sense of mission—preservation of Jewish survival—that community members feel by their religious status. This feeling bestows tremendous importance on educating the younger generation following the community's values and enhances the community's resilience (Marty and Appleby 2004).

Despite strong communal identity, Litvish groups are internally differentiated by their affiliation with a specific Yeshiva (higher institute of theological study). Litvish men from the whole range of Yeshivas maintain a standard dress code and outward appearance that distinguishes them from the external society. The curriculum of the Litvish schools places great emphasis upon Jewish studies. Anglo-Jewry possesses some Yeshivas conducted as part-time, full-time and boarding institutions. The most famous of these is in Gateshead, and there are others in London, Manchester and Sunderland. In what follows, the research will focus on two Litvish communities in the UK, in Golders Green and Gateshead, and will show how interrelation and intra-relations with other groups motivated each community to establish delineation between Halakhic obligations and the urban reality, which the urban environment shapes by. The status and influence of the Litvish and the difference between the lifestyles of the communities in London and Gateshead make the Litvish an interesting case study for analysing how group identity, individuals' preferences and implementation of communal norms moulds residential patterns.

The Litvish community holds religious study as a central cultural value. Nevertheless, life in modern urbanites, where most of Britain's Haredi population is concentrated, presents socio-economic and religious challenges. The Board of Deputies (2013) describes that "the philanthropists of Haredi society are also unable to continue supporting the growing number of Yeshiva scholars in London, particularly since they are being called upon to preserve the large society of learners in Israel, which is also struggling for survival." The accepted norm whereby married women devote themselves to the socialisation of the younger generation and housekeeping places the responsibility for making a living mostly on the men. In particular, individuals are expected to organise their lives in a way that maintains religious commitments and communal identity while adapting to modern reality. Increased familiarity and certain intimacy with the society at large and its particular sociocultural meaning codes, norms and values became a means of sustaining the Litvish group's particular against niche and structural position. The members of the Litvish community established their particular niche position within the general structure of the Torah society reappraise the appropriateness and functions of numerous sociocultural traits which had hitherto constituted part of the basis for their ethnic identity. As long as these adjustments did not interfere with religious precepts, they were acceptable, as people could remain Haredi. In so far as community members wish to retain specific components of their sociocultural commonalities, they also recognise the need to provide scope for cultural adaptations in areas of interaction with the general society. This is a social daily reality which they can neither avoid nor ignore while occupying the niche of ethnic hegemony. The next chapter exemplifies how this tendency caused the boundaries of the Litvish enclave of Golders green to become less rigid and in turn jeopardised selective pre-existing cultural commonalities.

References

Bilu Y (2003) From Milah (circumcision) to Milah (word): male identity and rituals of childhood in the Jewish ultra-orthodox community. Ethos 31(2):172–203

Etkes I, Tikochinski S (2004) Memoirs of the Lithuanian yeshiva

Lupo J (2004) Can Shas restore past glory?. The Floersheimer Institute for Policy Studies, Jerusalem (In Hebrew)

Marty ME, Appleby S (2004) The fundamentalism project. University of Chicago Press, Chicago

Stampfer S (2012) Lithuanian yeshivas of the nineteenth century: creating a tradition of learning. Oxford

Chapter 5
The Litvish Community of Golders Green: The Formation of Nested Residential Patterns

Abstract This chapter focuses on the relations between the social structure and the spatial patterns of Golders Green. By using mechanism-based explanations of a socio-spatial phenomenon, this chapter describes the entities and the activities (e.g. social norms and tendencies) both by themselves and in concert with organisational regularities and spatial patterns being observed (Machamer et al. 2000). The self-organising spontaneous enclave represented by the individual–community relations typifies Golders Green. As often occurs, the urban reality induces a gap between individual and group preferences and their implementation. Unlike other social enclaves, the Litvish community of Golders Green exhibits a relatively low level of segregation from the mixed hosting/charter general population (Flint Ashery 2012). The spontaneous enclave cannot create complete segregation from other groups. It is exposing the social currents typifying the community, with relatively high concentrations of various subgroups in certain "nests" of the neighbourhood. Similarly, spontaneous enclaves maintained by social-organisational and spatial mechanisms are likely to emerge in other religious enclaves. Collective behaviour, in this sense, is expressed by the very existence of few neighbours belonging to the Litvish community, which can create an isolated living environment and minimise the sense of unfamiliarity related to living among "others". From the perspective of urban dynamics, the Litvish community exemplifies the influence of the social relations within a micro-cosmos of a small community on the spatial organisation of dense urban fabric. The acceptance of the Litvish leadership by a broad swathe of the Haredi population is facilitating the accumulation of control over individuals' lives through social, organisational, and spatial mechanisms. In line with the theory, this research confirms that even a low level of social organisation can affect the segregated pattern of residential space. This case study, therefore, provides an opportunity to examine the relations between nuances of social organisation and the resulting voluntary congregations in "nested" places at various urban scales characterised by more intense social interactions.

Keywords Collective behaviour · Voluntary congregations · "Nested" places · Population distributions · Residential relations · Golders Green

S. Flint Ashery, *Spatial Behavior in Haredi Jewish Communities in Great Britain*, The Urban Book Series, https://doi.org/10.1007/978-3-030-25858-0_5

5.1 How Does Plural Urbanism Affect Religious Development?

Social and ethnoreligious enclaves, today part of the urban landscape throughout the world, are a central theme of urban studies. The creation and impact of ethnoreligious enclaves were empirically studied from various points of view, including those focusing on the hosting/charter population and the minority groups (Marcuse 1997). Much of the geographic literature deals with the segregation resulting from either inner forces, encouraging people to congregate as a means to preserve culture, language and customs (Boal 1996, 2008; Chivallon 2001; Kosmin and Levy 1981; Larmore 1990; Macedo 1995; Rawls 1993; Wahlstrom 2005) or external forces, creating spatial exclusion of unwanted populations from the majority's space (Boal 1978; Knox 1982; Kosmin and de Lange 1980; Lee 1977). The mode of differentiation depends vastly on the individuals' identity and their group's status—mainly relative to the "charter" dominant matrix into which new minority groups are inserted. The spatial dynamics of social enclaves of the ethnoreligious enclave and the role of individuals' preferences and social relations in shaping minority groups' space for strictly religious groups still awaits deeper insight (Waterman and Kosmin 1988; Valins 2003; Flint Ashery et al. 2010; Flint Ashery 2017).

Within the households economic constraints, residential patterns tend to result from free individual choices and householder migration (Ihlanfeldt and Scafidi 2002; MacIntyre 1984; MacIntyre and Macintyre 1988; Schelling 1971; Wahlstrom 2005), usually influenced by an individuals' views on their own identity and that of their neighbours (Dougherty and Huyser 2008; Semyonov and Tyree 1981). According to social identity theory (Tajfel 1972) and its updated version (Turner et al. 1987; Abrams and Hogg 1990), individuals' behaviour reflects identification with larger societal units. Individuals define themselves regarding their belonging to a social category, adopt its norms and lifestyle and eventually create a group (Hogg and Abrams 1988; Flint Ashery 2018c). In addition to other researches that examine the relationship between borders and identity formation, Beaugrand (2018) show that "separate identities are dependent on the existence of group categorisation, be they religious, cultural, economic, social or ethnic" (Kaplan and Hakli 2002; Meinhof and Galasiński 2002; Migdal 2004). The degree to which these concentrations form communities depends on the presence of a shared territory, significant social ties and meaningful social interactions (Hogg et al. 1995). Voluntary congregations of "friends"—members of the same or close groups—can preserve community identities, culture, language and customs (Boal 1996, 2008; Taylor et al. 1994; Knox and Pinch 2000). The intense awareness of Haredi people of religious identity motivates them to congregate in defined areas to tightly maintain their community and exclusively encounter people of the same values and affiliations (Boal 1978; Cimino 2011). In Giddens' (1984) words, "Self-identity is not a distinctive trait or even a collection of traits [...] possessed by the individual. It is the self as reflexively understood by the person regarding her or his biography" (Giddens 1984, 53). "People belonging to Haredi groups tend to limit their social contacts to people belonging to the same groups and subgroups, preferably inside the defined enclave … As individuals carefully select those to be trusted, they engage in the building of self-identity. Each group cooperates in its own religious and educational institutions and exclusive social networks and tends to compete with 'others' over spatial-cultural dominance, emphasising the strategic influence played by spatial homogeneity in inner-city areas and suburban districts" (see also Shilhav and Friedman 1985; Valins 2003).

Inner-city areas share common patterns of development; other development patterns are common in suburban districts. Both result from the spatial-historical constraints that typically operate in each type of space (Christensen and Hogen-Esch 2006; Iceland 2004; Knox and Pinch 2000; Telles 1995). Most suburbs spring up as the result of central planning and rapid occupation (Knox and Pinch 2000; Mobius and Rosenblat 2002). Government policies, real estate ownership and financial considerations contribute to the establishment of uniform residential patterns that affect individual and community preferences (McNair 2006). Conversely, most inner-city areas emerge as a result of local initiatives, subject to different opportunities and evolve at a slower pace and with more significant variation (Alfasi et al. 2013). As a result, suburban settlements have fairly homogeneous built qualities and populations, while inner cities are more heterogeneous and inhabited by a greater variety of groups. This does not apply, however, that the residents of the inner cities are more integrated than those populating the suburbs, though. This view is especially relevant when dealing with Haredi groups, whose social contacts are usually limited to the groups and subgroups (Shilhav and Friedman 1985). As each subgroup operates its own religious and educational institutions and exclusive social networks, the Haredi groups can be, in effect, highly segregated in both suburban and inner-city neighbourhoods (Valins 2003).

Over the years, the development of delineation between halakhic obligations and the real world has created an existential tension that threatens collective survival and identity. In particular, the need to make a living establishes a point of contact, comparison and competition with neighbouring groups, while at the same time impacting on the identity of individuals and their connection to the community. The concern lest communal identity be exposed to external influences leads members of the Haredi community to congregate and interact within a delineated physical and communal-spiritual sphere. Although the research could be relevant to other religious and perhaps to other traditional societies, this chapter focuses on the Litvish community—which is considered the most homogeneous Haredi group. As we will see shortly, the spatial encounter between the social norms and lifestyle of the Litvish society and the "other" communities of Golders Green shed light on how group identity, individuals' preferences and implementation of communal norms mould residential patterns in a super heterogeneous inner-city neighbourhood.

Starting in the seventeenth century, the undermining of man's traditional place in nature, the placing of his rights at the centre, and the quest for equality and social justice gave rise to a proliferation of values and world-views and presented a challenge to religious traditions (Hamilton 1995). Processes of secularisation and modernisation attracted populations to urban centres and civic pluralism led to competition between population groups. The following approaches usually explain the effect of these processes on religious development: According to the secularisation theory (Durkheim 1912/1964; Berger 1979), specialisation, social organisation and the quest for rationalism are the main factors which led to the decline of the control of religious institutions in society. Modernisation has led to a division of labour and specialisation of social institutions, some of which are self-organised. This development dissolves the traditional framework that moulded human experiences and replaces the common basis of morality and identity, to individualism; a transition from local-communal affiliation, where religion draws its power from the community and its representatives organise the lives of its members in a defined area to a pluralist diversity of meanings and experiences. Weber (1988) explained that when the individual carries the responsibility for choosing his institutions and his degree of affinity with them, the search for greater certainty led to an interpretation of religious ideas into economic behaviour. This view led to a rationalisation of the world of employment and a relinquishing of reliance on religion. When religions are forced to compete in the provision of traditions as a category of consumer products, by market logic, religious groups are then forced to choose between defensive strategies, to minimise their exposure to modernity and offensive strategy, through differentiation or extremism towards other groups.

Finke and Stark (1988), however, who examined co-operations between individuals and groups competing over space in industrialised towns in the USA, claim that urban pluralism boosts group identity by the challenge posed by others. According to this view, avoidance of the connection between religion and institutional elites, marketing different versions of religious faith to different audiences and responding to the needs of potential consumers, increases the competition between groups and therefore strengthens religious development. Finke and Stark (1988) coined the term "religious economies", arguing that just as in the free market economy model, and the absence of state oversight and religious monopoly, religious pluralism flourishes (Finke 1997, 2015). Rapid urban growth and the creation of a metropolis that is home to a range of groups spur each in its struggle to survive in the competitive market. When a variety of groups seek to respond to the needs of specific market segments, they compete among themselves, such that the level of specialisation, uniqueness and differentiation is augmented. This process leads to the growth of institutions that have been proposed as a replacement for religion, aiming to fill the vacuum left in the life of the individual. This competition encourages segregation intending to preserve the community and facilitates enlistment of the individual to prefer the group interest over his interest. Elements of both of these contrasting and complementary trends may be discerned in the indication of delineation between halakhic obligations and the reality and its implementation in the urban realm of Golders green. Throughout recent generations, processes of modernisation and secularisation led to an erosion of religion and religious tradition among most of Jewish society and came to be regarded as a threat to those seeking to remain committed to them, such as the Litvish Haredi Jews.

5.2 Residential Choice, Spatial Relations and the Formation of Residential Patterns

Urban residential patterns result from various levels of cooperation emerging between relatively free individuals (Wirth 1956). Indirect cooperation reflects similarities in the way people "read" and interpret urban space. Direct cooperation reflects economic interests and social organisation (Fischer 1982). Cooperation has been examined in models which exclude decision-making by individuals and explain conflicts between groups (Boal 2008). For example, urban ecologists of the early twentieth century suggested that "spatial segregation reflects individuals' preference to cooperate with others as a means of promoting their welfare" (Wirth 1956). The effect of cooperation between the city's free individuals on the urban structure was considered, for example, by the Chicago School at the beginning of the twentieth century. According to their "invasion–succession" theory (Park 1936; Hawley 1950) and its more updated explanation (Hawley 1950), "spatial competition between groups is an ecological process which enables us to describe collective behaviour". According to his view and "in the absence of clear behavioural norms, spontaneous social gatherings are the means to improve an individual's ability to cope with the challenges of urban life. As living costs influence these mechanisms, areas with different resources specialise in ways which give them a competitive advantage over other regions" (see also Blumer 1951). Over time the original population leave and the area changes.

Duncan and Lieberson (1959) argued that residential segregation is inversely related to assimilation, while Anderson (1962) claims that assimilation leads to non-sectoral spatial patterns. Peach (1975) explains that "the degrees of spatial similarity between groups are correlates and symbiotes of the level of social interaction between groups. The higher the degree of difference between the spatial distribution of groups within an urban area, the higher their social distance from each other. The correlate of complete spatial similarity is entire social interaction with high rates of intermarriage, while the correlation of segregation is the rejection, by the more senior group, of social interaction with the lower group" (Peach 1975, p. 1). Current models, however, claim that the individuals' tendency to cooperate is more complicated (Bastian et al. 2011). Modern, urban individuals are driven by the constant assessments of risks embedded in contacts with other people. As individuals carefully select those to be trusted, they engage in the building of self-identity. In Giddens' (1984) words, "Self-identity is not a distinctive trait or even a collection of traits […] possessed by the individual. It is the self as reflexively understood by the person concerning her or his biography" (Giddens 1984: 53): "The person's biography is, inevitably, connected to the urban sphere—again, in a complicated manner. Giddens theorises a time-space distinction which is typical of modern, post-traditional identities that do not derive from the particularities of localised spaces." Accordingly, post-traditional social relations highlight the role of social institutions at the expense of place. Individuals may be driven into segregating themselves from the influences of unwelcome groups and concentrating on what they perceive to be friendly social environments (Johnston et al. 2007). Previous researches show that "such voluntary congregations may create 'nested' places at various urban scales: buildings, street blocks, neighbourhoods and larger enclaves, with the smaller scales having more intense social interactions" (Kusenbach 2008). Urban segregation is therefore viewed as a multidimensional phenomenon where a person can reside among "others" but remain segregated regarding her or his identity as well as meaningful daily activities and interactions (Britton 2011).

Researchers who examined intergroup relations among Mormons, Catholics and Protestants (Finke 1992, 1997; Tausch et al. 2011; Phillips 2015) raise doubt regarding the degree to which a spatial segregation of post-traditional society is relevant for understanding the social life of strictly—religious groups residing in current multicultural cities (Bastian et al. 2011). Beyond customs and language differences, there is a structural mismatch between the tribal social order and the liberal approach that sees people as individuals. Alfasi (2014) describes that "belonging to a social group is, in most cases, innate rather than affected by personality and biography", and the process of transition from a tribe, to state, and then to European city "may not mean the demise of the previous tribal foundations, but rather the broadening of them" (see also Gellner 1990 and Flint Ashery 2018c). While several studies claim that ethnicity doesn't play an active role in residential preferences (Stillwell and Williams 2005; Catney and Simpson 2010), others have revealed that for ethnoreligious groups, belonging to a community is the primary source of strength and vitality. In the absence of regular status, kinship, friendship, fellowship and ethnoreligious ties become a form of "social capital". The intense awareness of identity by socio-religious adherents motivates them to congregate in defined areas to tightly maintain their community and exclusively encounter people of the same values and affiliations (Cimino 2011).

Robinson (1984) suggested a term—encapsulation—to explain "the spatial behaviour of a group in spatial concentration in a 'sort of social bubble', distinct from the host society, speaking a vernacular language, with their costume and rules of marriage". Vaughan and Penn (2006) explain how "such patterns of behaviour amongst relatively established immigrants … who have found that for some ethnic groups there is a continuing choice of clustering (albeit in a more dispersed pattern) to create opportunities for endogamy and sufficient population for religious and cultural institutions to thrive" (see also the explanation of Peach 1998). In the recent decades, the use of the notion of "melting pot" to describe different levels of residential integration between groups has been gradually replaced by the model of a "menudo chowder", where "most of the ingredients to melt, but some stubborn chunks are condemned merely to float" (Gomez-Pefna, in Bhabha 1994, pp. 218, 219). Building on this metaphor, Valins (2003) explains how boundaries, at least for several Jewish groups, are "not just temporary acts of resistance, but instead revolve around 'more stubborn' incommensurable elements that refuse (or are unable) to melt in the heat of a postmodern world."

Valins (2003) describes the ways in which post-colonial literature over the last decade has been "highly influential to geographical conceptions of identity and understandings of socio-spatial boundaries. Ideas of hybridity, interacculturation and in-betweenness have followed recognition that subjects are often positioned in-between or across domains of difference like race, class and gender, in the 'interstices' where these domains intersect" (see also Bhabha 1994; Watts 1997). Massey (1991a, b, c) explains how "with seemingly ever-increasing global interconnections, Cartesian-style notions of the bounded and stable nature of individual identities have been increasingly abandoned" (see also Pile and Thrift 1995; Hannerz 1999). As the "naturalness" of races, nations and the like are challenged, so understanding the movements and flows of cultural practices have become increasingly important (Lavie and Swedenburg 1996). Valins further suggests that "concentrating on movements and flows at the expense of stability and 'stubbornness' has concomitantly denied the grounded experiences of many individuals and groups who are actively seeking to resist the effects of global homogenisation and the loss of distinctive and traditional ways of life" (Smith 1993, 1999). The neat categories characterised by the turn to spatial sociology in the 1960s have been criticised by academics seeking to unpack the complexity and "fuzziness" of notions such as race and ethnicity.

5.3 The Jewish Settlement in Golders Green

Golders Green (Fig. 5.1) is located within the greater borough of Barnet in North-west London, close to the city centre. Although the settlement history of Golders Green goes back to the eighteenth century, the neighbourhood is essentially a late nineteenth-century suburban development, populated over a long period as a result of many bottom-up individual decisions. As part of London's urban processes, the traditional Jewish enclaves had expanded towards the borough of Barnet with the establishment of the Jewish cemetery adjacent to Hoop Lane in 1895. In the early twentieth century, Golders Green grew rapidly in response to the opening of a tube station of the London Underground railway (1907). With the availability of new suburban housing, Jews arrived in Golders Green in numbers. After Hitler's rise to power, the German Jewish immigrants took root and form the Golders Green Beth Hamedrash. Soon after, Galician Jewish immigrants built other synagogues and schools, and a prominent Jewish community has been established. Pam Fox (2018) describes how the Jewish identity and infrastructure evolved with the ability to buy Kosher bread locally in the 1930s and the establishment of the first Haredi Shtiebl (an informal place used for communal Jewish prayer. It is usually smaller than a synagogue and approached more casually) on 1932.

By the 1950s, the Jewish population in Golders Green tripled, and the neighbourhood gradually turned from a mixed neighbourhood to a neighbourhood that identified with the Jewish community. The success and sustainability of Golders Green as a vibrant Jewish area are attributed to the multi-layered Jewish infrastructure, community solidarity, adaptability to newcomers and a generally good relationship with the non-Jewish neighbourhood which has become increasingly multicultural (Fig. 5.2a, b). With different waves of Jewish immigrants—those from Germany and Austria, the Arab states, India, Egypt, Aden, Yemen, South Africa, Iraq and Israel, and, more recently, members of the Haredi community and Francophones—and with different religious groupings, the Jewish "community", really an amalgam of

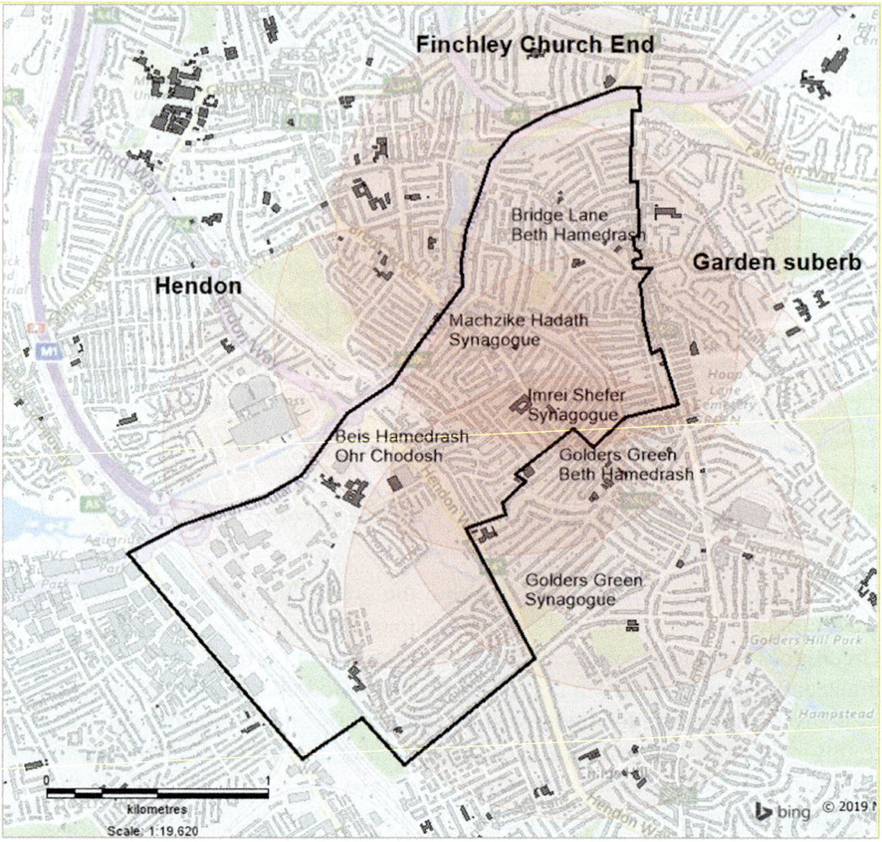

Fig. 5.1 Golders Green is located within the greater borough of Barnet in Northwest London

subcommunities, remained inherently ill-defined. Between 2001 and 2011, Golders Green's Jewish population increased by 34.6%, and according to the 2011 census, Jews made up 37% of the neighbourhoods' population. Today, Golders Green is the largest Jewish neighbourhood in the UK, with Christian, Muslim and people with no stated religion who represent the majority of Golders Green residents with significant people of Polish, Indian, Japanese and African origin. There are more than 40 synagogues dotted throughout the area, as well as 30 schools, many of them private. The neighbourhood is also the most significant Jewish Kosher hub in the UK (Fig. 5.3).

5.4 Golders Green's Free Communal Market

The residential choice and spatial activities are determined by socio-cultural-economic interactions and vice versa (MacIntyre and Macintyre 1988; Mobius and

(a)

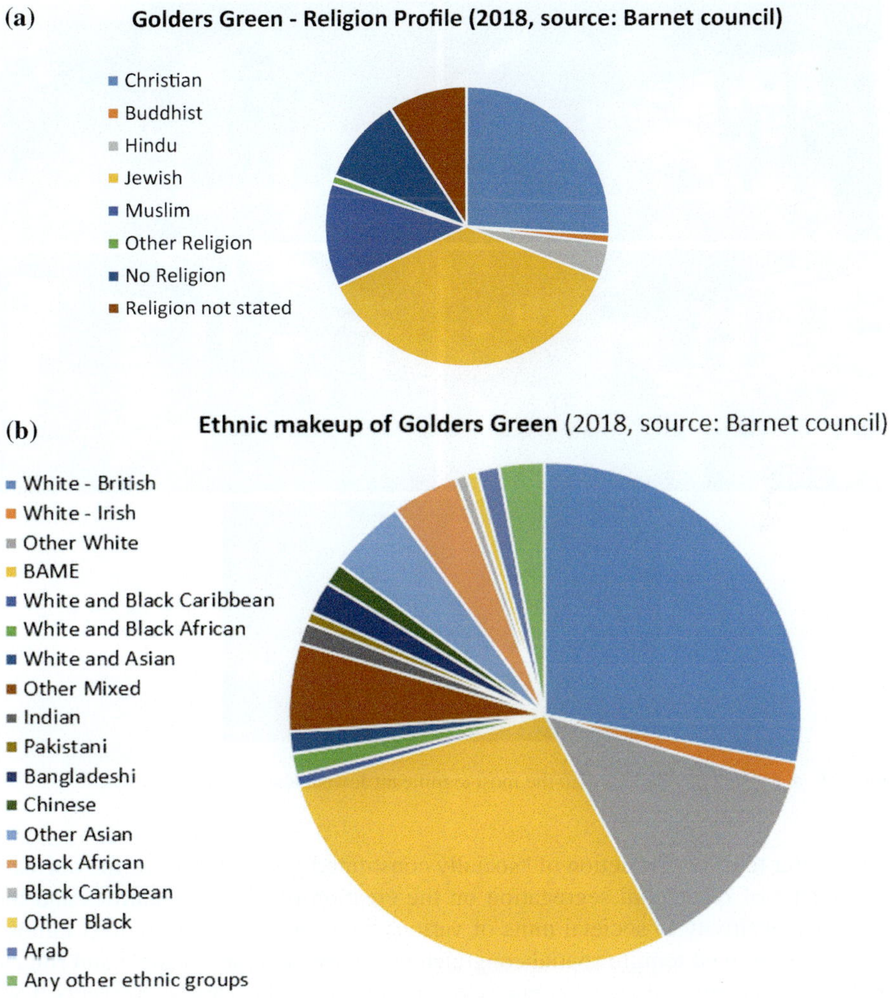

Fig. 5.2 **a** Religious and **b** ethnic profiles of Golders Green

Rosenblat 2002). The literature assumes that different levels of a social organisation play an important role in shaping segregated residential spaces (Christensen and Hogen-Esch 2006; Iceland 2004; Knox and Pinch 2000; McNair 2006; Mobius and Rosenblat 2002; Telles 1995). Previous study (Flint et al. 2012) explains how this viewpoint, in effect, "has a precursor in Greer's (1962) distinction between types of metropolitan communities: communities of space are local, usually suburban, for which the political reference is defined by people residing in a specific geographic area, whereas communities of interest have a larger metropolitan scope and are virtually a spatial". From a similar point of view, Schnell and Benjamini (2001, 2004,

Fig. 5.3 **a**–**e**: Golders Green is also the most significant Jewish Kosher hub in the UK

2005) refer to the construction of "socially constituted spaces reflecting the decreasing impact of residential segregation on the creation of current identities and the growing sensitivity to societal units of various types and scales. Religious minorities, however, who require spatial congregation for maintaining lifestyle and rituals, and whose meaningful social contacts are usually limited to their community, tend to combine spatial and social segregation" (Britton 2011).

Property prices in the Jewish areas of Golders Green are significantly higher than in similar areas (Fig. 5.4). This phenomenon is usually explained by the particular importance of housing in this sector and the need to reside within the community (Valins 2003; Schnell and Benjamini 2005; Flint Ashery 2018a). Golders Green's high housing prices (for both rental and purchase) reflect its desirable central location, the wide range of institutions active in the neighbourhood and the population's high status. The neighbourhood represents a unified housing market in which all residents are free to buy or rent apartments as they wish. This study, therefore, explores the residential patterns in Golders Green, traces the links between the social norms and the urban environment and reveals the significant mechanisms that determine the community's residential behaviour.

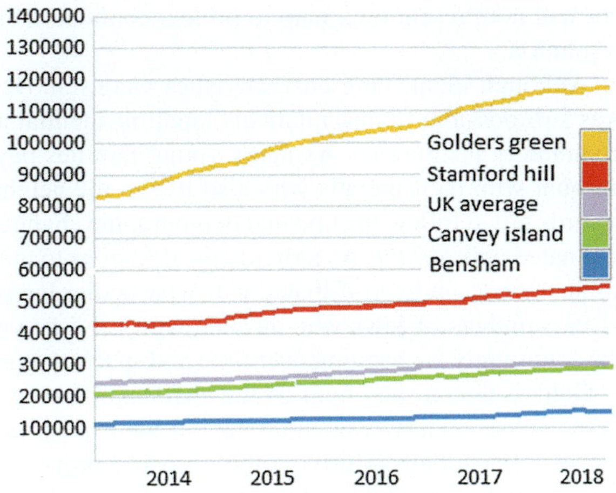

Fig. 5.4 Properties prices in the Jewish areas of Golders Green, Stamford Hill, Canvey Island and Bensham

5.5 Methodology: The Case Study of Golders Green

To explain the relations between ethnoreligious identity, residential relations and the spatial organisation of Golders Green, the method was framed around two approaches: the first approach follows the detailed residential patterns of Litvish households according to a synagogue membership survey that was took place between March and June 2017. As previously described, synagogues serve as communal centres, provide religious services and promote Jewish religious engagement with the broader Jewish cultural life. According to JPR (2017), 80.7% of Jews holds synagogue membership means that they are currently paying a membership fee to belong to an established synagogue. As such, synagogue membership figures provide a consistent indicator of patterns of Jewish affiliation over space and time and are regularly used to assess Jewish survey samples. As determining what a "synagogue" is, what does it mean to be affiliated with a particular synagogue and what is the exact nature of this association is becoming more complex, this research method will focus only on these six Litvish institutions (Fig. 5.1):

- Golders Green Beth Hamedrash
- Golders Green Synagogue
- Bridge Lane Beth Hamedrash
- Beis Hamedrash Ohr Chodosh
- Imrei Shefer Synagogue
- Machzike Hadath Synagogue.

The membership figures for the Litvish synagogues located in Golders Green were collected directly from synagogues to reduce the risk of double counting in this area at a level of five (NW11_0; NW11_8; and NW11_9) digits of postcode sector areas. In total, there are 1346 members by household in these synagogues). However, since most of the buildings are private houses, mainly typical British semi-detached, and the relatively small number of the Litvish occupant, the collected data were analysed but will not present on a map to protect confidentiality and keep the building-scale resolution.

Although identifying characteristics were removed from the spatial data set, it was still possible to link relatively small residential areas to synagogues and other community institutions. As many young families paying their fees through a joint account with their parents who also live in the neighbourhood, and sometimes the records fail to follow the updated demographic changes, the accuracy of the data was estimated only at 71%. As individuals' decisions to join a specific synagogue demonstrate their desire to be affiliated with a particular Jewish community, an examination of the numbers of Jews who affiliate with Litvish synagogues can indicate on how the community is changing over time and space. An in-depth analysis of these trends in Golders Green shed light on the challenges facing the community. Based on primary and secondary residential data of the Litvish households, this chapter reveals the social apparatus is driving the observed order in the Social and ethnoreligious Litvish enclave of Golders Green.

The second approach follows a "naturalistic" fashion in-depth interviews (Scheurich and McKenzie 2008). The questionnaires were shaped by an "interview guide" which indicated some key questions to clarify responses and the social processes that gave rise to spatial patterns. The interviews were "conducted as a conversation, with the specific content and length of the interview being flexible and variable across the respondents" (see Massey and Denton 1987). The interviewees are from different age groups (over 18) and most of them living in the neighbourhood. Twenty-two in-depth interviews were conducted with key figures of the Litvish community and outside the community to clarify responses and the social processes that gave rise to spatial patterns. Those involved with the communities were interviewed about spatial relations between the individual and the community and the economic aspect of the institutions regarding housing. Three rabbis and community leaders living in Golders Green provided information on the spatial relations between the individual and the community, the residential choice mechanisms and community life. Also, six residents from various "other" population groups were interviewed about activities in public and private spaces and their familiarity with the Litvish community, their appearance, characteristics and their spatial distribution. Finally, I spoke with four representatives of real estate agents from the community to verify the information obtained on residential mechanisms in Golders Green.

Data from the synagogues and the interviews were combined with aerial photographs and a geographic database provided by the ordnancesurvey.co.uk/opendatadownload/products.html (updated as of 2017). The data were then organised as a GIS layer, in which every record in the table is related to the corresponding postcode sector area. The geo-referenced data from the interviewees and the survey's spatial-temporal GIS enable evaluation of residential patterns at the resolution of a single postcode sector area. Compilation of the data produces information makes the micro-residential analysis of Golders Green's residents empirically possible. The information was crosschecked with blogs, articles, and Internet sites, which offered a range of different types of knowledge and perspectives on the communities.

5.6 Bottom-up Initiations: Communal Identity and Intergroup Residential Relations

Twenty-five out of the 27 interviewees who are affiliated with the local Litvish community attributed high salience to their neighbours' identity. Explaining that for them, the identity of their neighbours has the higher importance when they are choosing their new home (rather than the flat's price or the location of important institutions), they hint on the segregated attitudes held by the individuals. The responses also indicate the resemblance between individual preferences and the residential policy of the community leadership to stay in Golders Green and live within the community. Note that the interviewees declared on the segregated tendency even though Golders green is already identified with the Litvish community. With the respective trend confirmed, I attempted to determine the other socio-spatial mechanisms that led to the neighbourhood's residential pattern.

The case study of the Litvish Haredi population in Golders Green rise three fundamental questions: What is the impact of residential relations on the spatial organisation of Golders Green? How do residential patterns are determined by social, organisational and spatial mechanisms and vice versa? Would the residential space maintain its heterogeneity throughout the years, or segregated future is inevitable? To answer these questions, I, therefore, looked at the mechanisms that bridge the gaps between the individual preferences and the urban constraints based on primary and secondary residential data of the Litvish households. This chapter thus examines the backdrop stands the loose residential organisation characterising inner-city neighbourhoods and how individuals and the community implemented their housing preferences in the absence of a central community organisation. These mechanisms of communal identity and intergroup residential relations consist of three parts as described below.

5.7 Social Mechanisms

The constant tension between the modern lifestyle of the surrounding society and in particular the need for familiarity with core studies (e.g., the theory of evolution), non-Kosher food, liberalism and universal rights and the need to make a living within this society, leads to a range of combinations of religious study, education and employment. Avram (pseudonym, male, 43) explains: "I wanted to continue studying. I was a good student and when I was young everyone thought I could be a Rabbi or a Dayan. But it was vital for me to live in this neighbourhood and the prices here are very expensive because we all want to live here. The Yeshiva had no money to fund me because two others were better than me and they needed full funding to continue their studies. So I went to university. It was such a compromise then, and it was highly important for my wife that I will not stop learning (Torah). So I decided to study dentistry and orthodontics because hours are flexible and today I can combine Torah studies and work and live in the neighbourhood. This is the reason for me to be able to support my eldest son's studies in Jerusalem and my third son's studies in Gateshead."

The Litvish community of Golders green is recognised as "Yeshivish": An orthodoxy that is related to the Yeshiva institutes. Litvish women are generally married at age 18–20, after which they assume the exclusive responsibility for primary socialisation of the young children (Greenberg 1996). Rocha (pseudonym, female, 21) describes: "In the last year of seminary (high school for girls), everything changes. There is a sort of competition between the girls because they all want to get married. Marrying a scholar, of course. The boys are looking for girls from well-to-do homes whose parents can support their grooms' Torah studies. So it helps if you have Maatot—that you have a good character and so, but whose father can help a Torah scholar—it's really great for her. There were those who got married right after the seminary and then strolling around with a baby pram in the neighbourhood—so proud of themselves. Yeah, those who can stay in the neighbourhood can also be benefited from family supports, in addition to the community's".

Without financial support from the community, these social norms place the burden of supporting the family on the men. Accepting the Israeli-Litvish as the religious Ilit, young adults are sent to Israel or Gateshead to complete their studies. Litvish men are usually married in their early twenties (18–21), after spending a few years in Israeli's Yeshivas and continue full-time studies until the age of 28. Shlomo (pseudonym, male, 23) explains: "I studied for four years at Toras Moshe Yeshiva in Sanhedria Murchevet, Jerusalem. My brothers and my father also studied there; My father was one of the first student of this Yeshiva! We had started saving to enable our children to study there in the future as well. It was such a meaningful time, and I very much appreciate the family effort that helped me to do it. I used to live in the dormitories there. In addition to the Beis Medrash, they also have a basketball court! About a year ago I came back here to Golders Green. My wife was just about to finish the seminary when we got married. With the help of G-d, there is a child, and I work here teaching. It feels that the entire community is behind us." During this period, the Yeshiva is responsible for guiding the young couple in building the spiritual foundations of the home, and the young family enjoys the support of the broader family circle while maintaining the accepted norms and codes under the watchful eye of the community. Lea (pseudonym, female, 58) describes: "Get married and stay in the neighbourhood—I wish it to all of my grandchildren. Here everyone knows everyone, knows who to turn to and you get help before you even ask. It is such a great privilege to marry and bring children with a Talmid Hacham, to care for your children to grow up in the light of Torah and Maasim Tovim (good deeds)." Within this protective environment, the family undergoes self-organised processes that prepare it for fully independent living (Silber and Wertheimer 1992).

After finishing the fundamental period of religious study, most Litvish men choose to find their way into the professions when these occupations are open to them, and

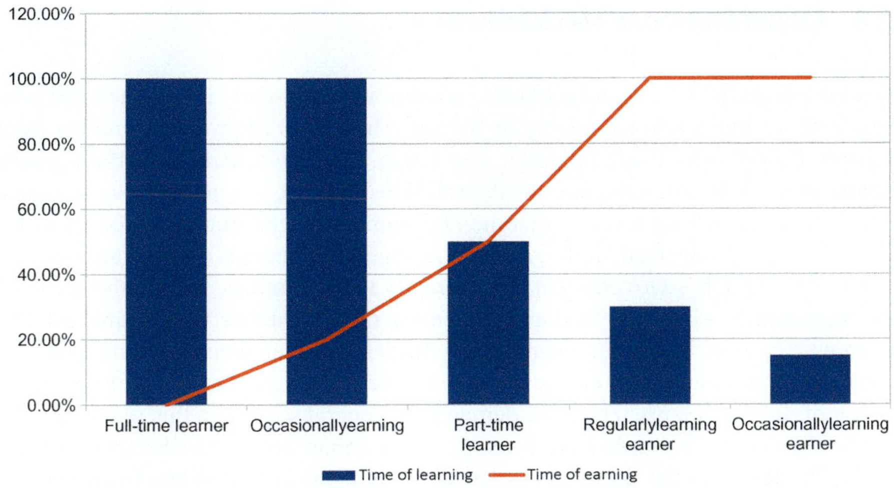

Fig. 5.5 Type of Haredi men by pattern of Torah learning and earning in the UK, according to Gonen's classification (2006)

opportunities for training are available. Universities attract Litvish men in considerable numbers, and many of them are studying medicine, law, and accountancy. Others are also engaged in research and teaching in the sciences and humanities. Following Gonen's intermediate model (Fig. 5.5), they considered as an "occasionally earning learners" or "part-time learners" or sometimes "regularly learning earners" (Gonen 2006). Rabbi Lev (pseudonym, male, 51) explains that at this stage, the responsibility for learning rests with the individual: "I had come from Manchester about ten years ago, and I can say that the daily life here is different because a person needs much more money to survive financially in Golders Green. We, the community's leaders, must encourage young men to study in universities and to acquire a profession. It becomes essential—otherwise, the community will not survive with the shrinking benefits. There will always be those who out marriage and leave, but the departure is not high because they know the right way. As a community we allow young men to get familiar with the outside world, to integrate with the general population up to a certain level. Usually, work-related. We are not Hasidic, you know, so when you get familiar with the outside world, there is no curiosity and much less desire to mix with these people. Our children can see that the differences are much higher than the similarities, and among our community, there is almost no departure. Given the attempt to merge the holy and the secular in their life, I really appreciate young adults (male) here work harder to overcome the high housing prices and live in the community." Although Gonen's definitions are also suitable for non-Litvish communities, it is typical mainly for the Litvish individuals to carry their responsibilities for choosing their institution and the degree of affinity with it. Each individual is expected to independently resolve the tensions between his communal obligations and his economic dependence on the external society, which offers a plethora of temptations.

5.8 Organisational Mechanisms

Over the years, historical and economic conditions have jointly established and consolidated a Litvish scholar society in Britain. The arrival of prominent rabbis from Eastern Europe who made this their new home, along with British welfare policy, contributed to the attractiveness of the religious lifestyle among Yeshiva students. Many families, whose level of income and number of children qualified them for government aid, received study grants, housing, child benefits and income supplements through UK's Government aid—directly to the families. Unlike the sectorial payment transfers in Israel, welfare payments in most Western countries are based on economic criteria. In Britain, the relatively few Haredi men who study full-time and meet the requirements that entitle them to a government grant are mere drops in the ocean of dole recipients. Thus, the demands of general society in this regard are not explicitly directed towards Haredi recipients (Berman 2000). Moshe (pseudonym, male, 38) describes the financial burden due to housing costs: "When I was young I grew up in this neighbourhood and it seemed natural to me to live within my community. But now, as a family man, I understand the importance of living here and the cost it requires, everything is so expansive now as the support of the state has been shrunk. The rent here is very high and I have four brothers. My parents saved so that we could take a mortgage and live in the neighbourhood. Now I save for my children so that they too can live in Golders Green. To live here is to be part of our community, to meet people everywhere, to be watched, to continue the tradition of generations." However, in the last decade, entitlement to a large portion of the government welfare payments has been conditional on joining the workforce, most of the Litvish men are integrated into the workforce, and the degree of benefits recipients has declined.

With the rise of individualism and secularism in the UK as elsewhere, religion becomes a choice that carries significant economic costs. Lea (pseudonym, female, 58) explains the budget priorities that involve large families: "Growing up in families like ours teaches you to manage your life responsibly with self-imposed criticism because the budget blanket is short. The most important thing is education, that's clear. For young people to be educated according to our values. This is the basics. If they learn well within the community, they will marry well. That's why we send to our schools. Where our own people work, where you can be sure that they are exposed to the right content and values from the earliest age. Health here the state finances and it helps a lot. Kosher food is expensive, and some places can offer large quantities, and this reduces costs. The children are also at school all day, and most of the time they eat there. Shabbat means a lot of expenses for large hospitality and disposable tableware. Weddings, for example, can be modest and be paid from the routine budget. For our children's flats, however, we are saving from the moment we get married." In the UK, the state covers school tuition costs for the national curriculum, but not religious studies. Hence, families who send their children to the community's educational institutions are committed to massive expenditure on private education. Other factors relate to the cost of living for such large families in central London and the costs of Kosher food.

Most Litvish men prefer occupations that allow them a degree of flexibility in their work schedule, allowing them to learn with study partners in between or after work hours, nearby the workplace or at home. The most popular occupations are teaching commerce and real estate. During the 2000s, there has been increasing interest in technological realms. Within the "full-time learner" category (Fig. 5.5), a distinction has developed between the "Ilui"—a young man with exceptional study abilities, who is supported through community or Yeshiva philanthropy, and the "Talmid Hakham"—a student supported by his family, or one who is willing to live in economic distress for the sake of devoting all his time to Torah. Full-time learners enjoy a high social status, which might also bring a "good match"—meaning, a bride from a family of respected lineage or with financial means. Full-time learning for one's entire adult life—the Litvish model in Israel—is not common, owing to the limited amount of learners who can receive a stipend from the Yeshiva and the heavy costs associated with funding a Talmid Hakham. At the same time, it is not considered acceptable to relegate religious study to the Sabbath and festivals alone, and the "occasionally learning earner" status carries a high social price. Today, about 20% of Litvish men in London devote themselves to full-time religious study, while about 70% combine life-time education with making a living.

In recent years, narrow interests and succession wars and extremism have led to conflicts of interests between the Litvish rabbinical and political arms of the Yeshivas. These tensions have torn the camp apart, weakening group discipline and the ties between the rabbinical leadership and the Yeshiva students. Examples of weakening discipline include increased Litvish individuals' involvement in education and employment frameworks. Rabbi Aaron (pseudonym, male, 64) describes how, as the economic power and involvement of the leadership in the community's daily life is limited, the local institutions focus mainly on their advisory role: "I see my role as responsible for our community—its cohesion and continuity. It is a complex role of advising and directing—obedience to the Rabbi is not obligatory in our community—and the degree of involvement in the community's life depends on each. If a person chooses to contribute less or even leave the Litvisher world, I cannot command or enforce, of course, only to talk to his heart and try to convince him. It's never pointless. If a family, for example, wants us to talk with a person who wanted to marry outside, this is Dinei Nefashot (Laws of persons), so we try to persuade him so that he returns to his place." Although the local rabbis of the Litvish community and the connection to a specific rabbi/Yeshiva—is quite weak and flexible, the local institutions accompany the individual throughout all stages of life are denounced. In return for annual membership fees paid directly to the synagogue (approx. £800 per family), they are providing a sense of security and belonging through life circle. The synagogue creates a common framework for the men who gather there for study and prayer after work hours and a social envelope for the women during the day. Religious lectures and cultural activities are also held at the synagogue, enhancing communal cohesion.

5.9 Spatial Mechanisms

5.9.1 The Formation of Nesting Congregation Area

Despite the weakening of the rabbinical leadership and the limited support, group identity is a significant element of the "sense of place" that binds a person to the communal environment. This fundamental human experience is expressed in the culture that develops within the enclave and is maintained, despite the rather limited involvement of the leadership in everyday life of the community and its weakening ability to enforce group discipline (Stadler 2002; Salmon et al. 2006). Weak community discipline, high living costs and a competitive housing market lead to more flexible housing choices and spontaneous social gathering of the Litvish community in their parochial realm. As such, young families tend to overcrowd in the designated area of the community. Construction irregularities are relatively minor. The emergence of "nested" places—smaller communal areas inside the Litvish area of Golders Green—is therefore likely to occur.

Despite the high housing costs in Haredi areas of Golders Green and the variety of population groups, communal identity and cohesion strengthen this group behaviour and create a sharp contrast to the prevailing social organisation in the wider urban sphere. Nesting congregation while integrating into the general population enables the Litvish group to improve their ability to cope with local challenges. Previous studies coined the term 'micro-segregation' to descrive the dynamics of this spatial process (Flint Ashery et al. 2012; Flint Ashery 2018b). An example of such a definite order is a dense inner city area of Golders Green. Following the synagogue membership data, Fig. 5.1 illustrates the tendency of families identified with the Litvish community to live in some regions of the neighbourhood. In addition, analysing the data indicates on a clear preference of the community for living on the inner streets of the neighbourhood than on the main roads and suggests on the nests of segregation along the inner streets of the neighbourhood. In these inner streets, the level of noise and pollution is lower compared to the main roads, and the living condition is favourite by families with children. These nests are characterised by some 12–18 families (dwellings are usually in private homes) located within an estimated 20 min walk from the synagogues, to enable them to obey the laws of the Sabbath (which prohibit driving or riding in a car or any other vehicle from sundown on Friday until sundown on Saturday).

Relatively free housing choices and spontaneous social gathering within the community enclave enable the Litvish community of Golders Green to maintain its identity and rely on communal institutions. The tendency to send young people for long periods of training in Gateshead or Jerusalem is also attested to the Litvish trend to be affiliated with the community institution (in contrast to the rabbinical affiliation typical of Hasidim). The tendency to send the youngsters to live in the Torah centres reduces the community's pressure on the space. Naturally, the fear that the older population remaining at Golders Green will integrate into the general society is smaller. Unbounded by communal directives, the residential choices of Golders Green's inhabitants seem to reflect the state of the housing market and individual preferences to create an overtly integrated residential pattern. In addition to the Litvish individuals' involvement in education and employment frameworks, another example of weakening discipline of the Litvish rabbinical is the gradual acceptance of new cities, such as Westcliff, as a possible settlement detached from the traditional enclave.

5.9.2 Marco-residential Relations

The tolerance to "very different" groups—the range of ethnic groups that are drawn to the area—increases its heterogeneity and strengthens the Litvish communal cohesion. Other groups living in these areas—mainly the Japanese, the Turks and the Indians—are culturally different and hence present no threat to the Litvish way of life. Rachel Forman, a Litvish resident of Golders Green, explained: "I describe to my child the difference between them and us, and that strengthens him in our path" (Interview, 27 February 2012). Lesley Soan, a Tokyo-born resident of Golders Green who has lived next door to a Litvish family for the past seven years, explains his decision: "Jews make good neighbours. They have no alcoholism or violence. Even the recent [summer of 2011] riots didn't involve the Jewish areas. It is much safer here. That is the reason for people to want to live here, and the reason for prices in this area to be so high." Antony Spencer, another resident from the Litvish community, adds: "They're very different from us in every way: they look different, dress differently, speak differently. Our times are different from theirs. It's easier for us to live with them than with Reform, Liberal or traditional [Jews]. Certainly more than with Hasidim, Central orthodox, Reform or secular Israelis [Jews]" (Interview, 1 March 2012). Spencer's interview hints that when the differences are clear and significant, the threat entailed in exposing the younger generation to the lifestyles of other population groups is decline.

Despite the high housing costs in Golders Green and the variety of population groups, communal identity and cohesion strengthen this group behaviour and create a contrast to the prevailing social organisation in the wider urban sphere. The individuals' preference of the Litvish community to cooperate with others is a resemblance to the collective behaviour, identified by the Chicago School. The existence of other Jewish groups as well as other non-Jewish groups is what drives the Litvish community's processes of segregation, preferring to live in a micro-scale of nesting segregation. The need to make a living, as well as the consumer culture, erodes the time set aside for Torah study. Individuals organise themselves independently, such that the delineated area of the Litvish communal space is becoming more flexible, influenced and moulded by the penetration of "very different" groups. This process, where the individuals' wealth maintains the community congregation, began in Hackney in central London, continued to Edgware, and is currently moving towards north-western London, with an attendant rise in housing price. Adam Boyd (ps), one of 21 interviewees who describe this process, explains that the Litvish community can bear heterogeneity up to a certain limit; otherwise, they prefer to move. "We know such processes from Edgware, for example, where a non-Jewish population slowly entered the neighbourhood. Properties' prices fell, and the area becomes very heterogeneous, too much heterogeneous. And then we felt uncomfortable to walk in the street, sing in Fridays evenings before Sabbath and sometimes there are not enough people for a minyan (the quorum of ten Jewish adults required for certain religious obligations). So, again, we had to leave the neighbourhood". The community leadership, fearing that having individual families moving to new areas will threaten community cohesion, wanted to moderate the crisis by having families migrate as a group to a new area. However, this organised group behaviour is conflicting with the urban reality that requires members of the community to integrate and engage in day-to-day contact with the processes of modernisation. Previous attempts to emulate the moving to the suburbs of Haredi groups in Israel and New York failed in Shenley (during the beginning of the 1980s) and in the city of Milton Keynes (at the beginning of 2005). A unique case study is the scholar community of Gateshead that will discuss in the next chapter.

5.10 Discussion: The "Menudo Chowder" Litvish Community of Golders Green

This study revealed new aspects of the relations between the social structure and the spatial patterns of Golders Green. Analysing the spatial dynamics in the neighbourhood, where the housing prices are relatively high and the impact of the relations between individuals' and community leaders are relatively low, the research revealed micro-scale, nested pattern of residential segregation within a highly heterogeneous inner-city neighbourhood. In the absence of central planning and authoritative guiding leadership, the spatial structure of the spontaneous enclave is the creation of individuals.

Despite the interviewees' stated preferences to reside among "friends", the Litvish community of Golders Green live in a heterogeneous environment. In this regards, Kosmin et al. (1987) note that "the continuation of clustering among British Jews (in many cases third- or fourth-generation descendants of immigrants), yet show that they avoid any degree of spatial segregation which might isolate them from the benefits of fully functional integration into society." It, therefore, appears that the local residential structure of the majority of households is an insignificant dissonance with their individual needs and desires. These findings make it difficult to understand how Golders Green and other "unplanned" Haredi core neighbourhoods, such as Hendon, which exhibit similar internal organisation, have sustained such high prestige while also being so heterogeneous. As noted, Golders Green is a central and attractive residential neighbourhood, as indicated by its high housing prices.

Within the neighbourhood, the Litvish community tend to congregate in certain inner areas and to create for themselves conditions which are less than those of the "ghetto" and more than those of the ordinary environment. The synagogues, Jewish voluntary associations, Kosher butcher shops and Jewish groceries are not the centres of compact Jewish sectors, but rather the nuclei of Jewish populations which live interspersed with non-Jewish neighbours. Golders Green in this sense takes on a decidedly Jewish flavour, but they are not large Jewish quarters in the same way as the Litvish area of Gateshead, the former area of Whitechapel (East-London) and some districts of the provincial cities were once Jewish enclaves.

From the perspective of urban dynamics, the Litvish community exemplifies the influence of the social relations within a micro-cosmos of a small community on the spatial organisation of dense urban fabric. In line with the theory, this research confirms that even a low level of social organisation can affect the segregated pattern of residential space. This unique study's cases, therefore, provide an opportunity to examine the relations between nuances of social organisation and the resulting voluntary congregations in "nested" places at various urban scales characterised by more intense social interactions. Collective behaviour, in this sense, is expressed by the very existence of few neighbours belonging to the Litvish community that has the ability to live with members of the community while also minimising the sense of unfamiliarity related to living among "others".

Residential patterns in urban spaces express the compromises of individuals versus urban constraints and personal abilities. As often occurs, the urban reality induced a gap between individual and group preferences and their implementation. Similar to other social enclaves, the Litvish community of Golders Green exhibit segregation from the mixed hosting/charter general population. The spontaneous enclave lacks the ability to create complete segregation from other groups. It is exposing the social currents typifying the community, with relatively high concentrations of various groups in certain "nests" of the neighbourhood. Although this spontaneous micro-enclave maintained by relatively low social-organisational and spatial mechanisms are likely to emerge in other religious enclaves, it is not the case of Gateshead, the destination of many of the offspring of Golders Green's Litvish, as we will see in the next chapter.

References

Abrams D, Hogg MA (1990) Social identification, self categorization and social influence. Eur Rev Soc Psychol 1(1):195–228

Alfasi N (2014). Doomed to informality: Familial versus modern planning in Arab towns in Israel. Planning Theory & Practice, 15(2), 170-186.

Alfasi N, Flint Ashery S, Benenson I (2013) Between the individual and the community: residential patterns of the Haredi population in Jerusalem. Int J Urban Reg Res 37(6):2152–2176

Anderson TR (1962) Social and economic factors affecting the location of residential neighborhoods. Pap Reg Sci 9(1):161–170

Bastian B, Loughnan S, Koval P (2011) Essentialist beliefs predict automatic motor-responses to social categories. Group Proc Intergroup Relat 14(4):559–567

Beaugrand C (2018) Borders and spatial imaginaries in the Kuwaiti identity. Geopolitics 23(3):544–564

Berger CR (1979) Beyond initial interaction: uncertainty, understanding, and the development of interpersonal relationships. In: Giles H, St. Clair RN (eds) Language and social psychology, pp 122–144. Blackwell, Oxford

Berman E (2000) Sect, subsidy, and sacrifice: an economist's view of ultra-orthodox Jews. Q J Econ 115(3):905–953

Bhabha HK (1994) The postcolonial and the postmodern. In: The location of culture, p 172

Blumer H (1951) Collective behavior. In: Lee AM (ed) New outline of the principles of sociology. Barnes and Noble, New York, pp 166–222

Boal FW (1978) Territoriality on the Shankill-Falls divide, Belfast: the perspective from 1976. In: An invitation to geography, pp 58–77
Boal FW (1996) Integration and division: sharing and segregating in Belfast. Plan Pract Res 11(2):151–158
Boal FW (2008) Territoriality on the Shankill-Falls divide, Belfast. Ir Geogr 41(3):349–366
Britton ML (2011) Close together but worlds apart? Residential integration and interethnic friendship in Houston. City Community 10(2):182–204
Catney G, Simpson L (2010) Settlement area migration in England and Wales: assessing evidence for a social gradient. Trans Inst Br Geogr 35(4):571–584
Chivallon C (2001) Religion as space for the expression of Caribbean identity in the United Kingdom. Environ Plan D Soc Space 19(4):461–483
Christensen T, Hogen-Esch T (2006) Local politics. ME Sharpe
Cimino R (2011) Neighborhoods, niches, and networks: the religious ecology of gentrification. City Community 10(2):157–181
Dougherty KD, Huyser KR (2008) Racially diverse congregations: organizational identity and the accommodation of differences. J Sci Study Relig 47(1):23–44
Duncan OD, Lieberson S (1959) Ethnic segregation and assimilation. Am J Sociol 64(4):364–374
Durkheim E (1912/1964) The elementary forms of the religious life. Allen & Unwin, London
Finke R (1992) An unsecular America. In: Bruce S (ed) Religion and modernization: sociologists and historians debate the secularization thesis. Clarendon Press, Oxford, pp 145–169
Finke R (1997) The consequences of religious competition: supply-side explanations for religious change. In: Young LA (ed) Rational choice theory and religion: summary and assessment. Routledge, NewYork, NY
Finke R (2015) Religion and struggle in the European Union: confessional culture and the limits of integration. In: Nelsen BF, Guth JL. Georgetown University Press, Washington, DC
Finke R, Stark R (1988) Religious choice and competition. Am Sociol Rev 63(5):761–766
Fischer CS (1982) To dwell among friends: personal networks in town and city. University of Chicago Press, Chicago
Flint Ashery S (2012) The economic aspect of communal identity: an examination of the Lithuanian community in Jerusalem and London, Tabur, Resling and the Richard Koebner Centre, Jerusalem 5, pp 124–195 (In Hebrew)
Flint Ashery S (2017) 'Decision not to decide': a new challenge for planning. Euro Plan Stud 25(6):1076–1098
Flint Ashery S (2018a) Schelling-type micro-segregation in a Hasidic enclave of Stamford-Hill. Hous Stud 33(7):1038–1059
Flint Ashery S (2018b) Horizontal examination: micro-segregation mechanism in the diverse area of Whitechapel. In: Micro-residential dynamics, pp 25–34. Springer, Cham
Flint Ashery S (2018c) Effects of identity and belonging on residential experiences, perceptions and practices of the undocumented population in Whitechapel. In: Micro-residential dynamics, pp 73–94. Springer, Cham
Flint Ashery S, Alfasi N, Benenson I (2010) Micro-segregation of Haredim in Jerusalem. Israeli Sociol 12:81–110 (In Hebrew)
Flint Ashery S, Benenson I, Alfasi N, Bekman Y (2012) Schelling-like residential dynamics in a Haredi neighbourhood in Jerusalem. In: Diappi L (ed) Emergent phenomena in housing markets. Springer-Verlag, Berlin Heidelberg
Flint S, Benenson I, Alfasi N (2012) Between friends and strangers: micro-segregation in a Haredi neighborhood in Jerusalem. City Community 11(2):171–197
Fox P (2018) The Jewish community of golders green, a social history, 0750965878
Gellner E (1990) Tribalism and the state in the Middle East. In: Khoury PS, Kostiner J (eds) Tribalism and the state in the Middle East. University of California Press, Los Angeles, pp 109–126
Giddens A (1984) The constitution of society: outline of the theory of structuration. University of California Press, Berkeley and Los Angeles
Gonen A (2006) Between Torah learning and wage earning. The London experience and lessons for Israel. The Floersheimer Institute for Policy Studies, Jerusalem (in Hebrew)
Greenberg B (1996) Ultra-orthodox women confront feminism. Moment 21(3):36–38

Greer SA (1962) The emerging city: myth and reality. Transaction Publishers
Hamilton M (1995) The sociology of religion: theoretical and comparative perspectives. Routledge, London
Hannerz U (1999) Reflections on varieties of culturespeak. Eur J Cult Stud 2(3):393–407
Hawley F (1950) Big kivas, little kivas, and moiety houses in historical reconstruction. Southwest J Anthropol 6(3):286–302
Hogg MA, Abrams D (1988) Social identifications
Hogg MA, Terry DJ, White KM (1995) A tale of two theories: a critical comparison of identity theory with social identity theory. Soc Psychol Q 255–269
Iceland J (2004) Beyond black and white: metropolitan residential segregation in multi-ethnic America. Soc Sci Res 33(2):248–271
Ihlanfeldt KR, Scafidi B (2002) Black self-segregation as a cause of housing segregation: evidence from the multi-city study of urban inequality. J Urban Econ 51(2):366–390
Johnston R, Poulsen M, Forrest J (2007) The geography of ethnic residential segregation: a comparative study of five countries. Ann Assoc Am Geogr 97(4):713–738
Kaplan DH, Häkli J (eds) (2002) Boundaries and place: European borderlands in geographical context. SIAM
Knox P (1982) Urban social geography: an introduction. Longman, London
Knox P, Pinch S (2000) Bodies, sexuality and the city. Urban Soc Geogr
Kosmin BA, De Lange DJ (1980) Conflicting urban ideologies: London's new towns and the metropolitan preference of London's Jews. Lond J 6(2):162–175
Kosmin BA, Levy C (1981) The work and employment of suburban Jews, 6. Board of Deputies Local Study
Kosmin BA, Ritterband P, Scheckner J (1987) Jewish population in the United States, 1986. In: The American Jewish year book, pp 164–191
Kusenbach M (2008) A hierarchy of urban communities: observations on the nested character of place. City Community 7(3):225–249
Larmore C (1990) Political liberalism. Polit Theory 18(3):339–360
Lavie S, Swedenburg T (eds) (1996) Displacement, diaspora, and geographies of identity. Duke University Press
Lee TR (1977) Race and residence. Oxford University Press, Oxford
Macedo S (1995) Liberal civic education and religious fundamentalism: the case of God v. John Rawls? Ethics 105:468–496
Machamer P, Darden L, Craver CF (2000) Thinking about mechanisms. Philos Sci 67(1):1–25
MacIntyre A (1984) After virtue: a study in moral theory. University of Notre Dame Press, Notre Dame, IN
MacIntyre AC, Macintyre AC (1988) Whose justice? Which rationality?, vol 239. Duckworth, London
Marcuse P (1997) The enclave, the citadel, and the ghetto: what has changed in the post-Fordist US city. Urban Aff Rev 33(2):228–264
Massey D (1991a) The political place of locality studies. Environ Plan A 23(2):267–281
Massey D (1991b) A global sense of place. Marxism Today, June
Massey D (1991c) Flexible sexism. Environ Plann D Soc Space 9(1):31–58
Massey DS, Denton NA (1987) Trends in the residential segregation of Blacks, Hispanics, and Asians: 1970–1980. Am Sociol Rev 802–825
McNair B (2006) Cultural chaos: journalism and power in a globalised world. Routledge
Meinhof UH, Galasiński D (2002) Reconfiguring East-West identities: cross-generational discourses in German and Polish border communities. J Ethn Migr Stud 28(1):63–82
Migdal JS (ed) (2004) Boundaries and belonging: states and societies in the struggle to shape identities and local practices. Cambridge University Press
Mobius M, Rosenblat T (2002) The process of ghetto formation: evidence from Chicago. Manuscript, Harvard University
Park RE (1936) Succession, an ecological concept. Am Sociol Rev 1:171–179
Peach C (1975) Urban social segregation. Longman Publishing Group

Peach C (1998) South Asian and Caribbean ethnic minority housing choice in Britain. Urban Stud 35(10):1657–1680
Phillips D (2015) Claiming spaces: British Muslim negotiations of urban citizenship in an era of new migration. Trans Inst Br Geogr 40(1):62–74. https://doi.org/10.1111/tran.12054
Pile S, Thrift NJ (eds) (1995) Mapping the subject: geographies of cultural transformation. Psychology Press
Rawls J (1993) The law of peoples. Crit Inq 20(1):36–68
Robinson V (1984) Asians in Britain: a study in encapsulation and marginality. In: Geography and ethnic pluralism, pp 231–257
Salmon Y, Ravitzky A, Ferzinger A (2006) Orthodox Judaism: new perspectives
Schelling TC (1971) Dynamic models of segregation. J Math Sociol 1(2):143–186
Scheurich J, McKenzie K (2008) Foucault's methodologies. In: Collecting and interpreting qualitative materials, vol 3, pp 313–349
Schnell I, Benjamini Y (2001) The sociospatial isolation of agents in everyday life spaces as an aspect of segregation. Ann Assoc Am Geogr 91(4):622–636
Schnell I, Benjamini Y (2004) From Chicago and Los-Angeles to Tel-Aviv-Jaffa: a social areas model. Horiz Geogr 60/61:29–40
Schnell I, Benjamini Y (2005) Globalisation and the structure of urban social space: the lesson from Tel Aviv. Urban Stud 42(13):2489–2510
Semyonov M, Tyree A (1981) Community segregation and the costs of ethnic subordination. Soc Forces 59(3):649–666
Shilhav Y, Friedman M (1985) Growth and segregation: the ultra-orthodox community of Jerusalem. The Jerusalem Institute for Israel Studies, Jerusalem (in Hebrew)
Silber MK, Wertheimer J (1992) The uses of tradition: Jewish continuity in the modern era
Smith SJ (1993) Bounding the borders: claiming space and making place in rural Scotland. Trans Inst Br Geogr 291–308
Smith SJ (1999) The cultural politics of difference. In: Human geography today, pp 129–150
Stadler N (2002) Is profane work an obstacle to salvation: the case of ultra-orthodox Jews in Israel. Sociol Relig 64(4):455–474
Stillwell J, Duke-Williams O (2005) Ethnic population distribution, immigration and internal migration in Britain: what evidence of linkage at the district scale. In: British society for population studies annual conference, pp 12–14, University of Kent, Canterbury, Sept
Tajfel H (1972) Some developments in European social psychology. Eur J Soc Psychol 2(3):307–321
Tausch N, Hewstone M, Schmid K, Hughes J, Cairns E (2011) Extended contact effects as a function of closeness of relationship with ingroup contacts. Group Proc Intergroup Relat 14(2):239–254
Taylor DM, Moghaddam FM (1994) Theories of intergroup relations: international social psychological perspectives, 2nd edn. Praeger, New York, NY
Telles EE (1995) Structural sources of socioeconomic segregation in Brazilian metropolitan areas. Am J Sociol 100(5):1199–1223
Turner JC, Hogg MA, Oakes PJ, Reicher SD, Wetherell MS (1987) Rediscovering the social group: a self-categorization theory. Basil Blackwell
Valins O (2003) Defending identities or segregating communities? Faith-based schooling and the UK Jewish community. Geoforum 34(2):235–247
Vaughan L, Penn A (2006) Jewish immigrant settlement patterns in Manchester and Leeds 1881. Urban Stud 43(3):653–671
Wahlstrom AK (2005) Liberal democracies and encompassing religious communities: a defense of autonomy and accommodation. J Soc Philos 36(1):31–48
Waterman S, Kosmin BA (1988) Residential patterns and processes: a study of Jews in three London boroughs. Trans Inst Br Geogr 79–95
Watts EK (1997) An exploration of spectacular consumption: Gangsta rap as cultural commodity. Commun Stud 48(1):42–58
Weber M (1988) Gesammelte Aufsätze. 7 Bände. UTB/Mohr-Siebeck, Tübingen. ISBN 3-8252-1496-6
Wirth L (1956) Community life and social policy. University of Chicago Press, Chicago

Chapter 6
The Litvish Community of Gateshead: Reshaping the Territoriality of the Neighbourhood

Abstract The renewed interest and deep concern with "Torah Lishmo" [for His sake, for a higher, spiritual purpose] and the specific interest in Mussar [a Jewish spiritual, ethical, educational and cultural practice that gives concrete instructions on how to live a meaningful and ethical life] have created a community whose main concern is that of living lives which are in meticulous compliance with Torah precepts and with particular attention given to interpersonal ethics. Although there are other Haredi communities in Britain, Gateshead's particular role within the world of Jewry and its relative independence from the society at large provides the community with a specific position. It is an acquired position which is based upon a combination of their reputation as one of the world's most important centres of formal Jewish culture transmission and the homogeneity and level which the community demonstrates regarding their deep-felt religiosity and devotion to Jewish ethics. It would seem that Gateshead has in some ways become a yardstick with which members of other Jewish communities might evaluate their levels of adherence to, as well as a choice of sociocultural commonalities upon which they base their ethnic identity. Throughout fifty years, these changes had restyled interpersonal interactions and brought about a renegotiation of particular sociocultural constructs. The spatial division in the voluntary ghetto of Gateshead allows religiously based lives to remain cut off from external influences and safeguards the younger generation from the perceived threats of non-Litvish culture. In terrain where sectoral interests clash, they turn to unofficial modes of group action to achieve their goals. This tendency is a strategy which has been possible to accomplish due to the community's relative economic independence from the society at large and its occupation of a specific economic niche within the larger ethnic group, but it has also become desirable due to a particular sociocultural response to internal communal forces. Gateshead subgroup position is founded on the community's particular response to external and internal forces, and whose present ethnic communalities are expressed in ways which differ from other Jewish subgroups. It provides a subgroup position which is both recognised and pronounced by members as well as non-members of the community and could thus be specifically relevant for the residential patterns of tightly bonded religious groups and national/ideological minorities. The tangible outcome of these changes is that the community is not only territoriality distinguishes itself within Anglo-Jewry, but has also become increasingly isolated from the society at large.

S. Flint Ashery, *Spatial Behavior in Haredi Jewish Communities in Great Britain*, The Urban Book Series, https://doi.org/10.1007/978-3-030-25858-0_6

Keywords Group behaviour · Cultural dominance · Territoriality · Voluntary ghetto · Residential relations · Gateshead

6.1 The Creation of Cultural Dominance Within the Defined Enclave

Central to the contemporary study of delineation is the action of indicating the exact position of a border or boundary, including the process of bordering. As a political act, it is also a demarcation of difference that helps define and code the "border terminologies which focus on the binary distinctions between the 'us' and 'them', the 'included' and the 'excluded'" (Newman 2006), what is "same" and what is "other". As Pratt (1999) discusses, "the drawing of lines can, in addition to being an act of control by the powerful, also be an act of resistance by the weak". Alongside the creation of cultural dominance within the defined enclave, "the isolated territories serve as a protective niche, enabling conservation of lifestyles and cultures" (Komarova 2008). This behaviour, characterising organised groups such as group purchasing organisations and gated communities, is particularly strong among traditional communities (Bankston and Zhou 1995; Flint Ashery et al. 2014). Many of these communities' "attempt to revive traditional lifestyles by reinforcing compliance due to identification (as opposed to compliance out of fear or under explicit threat). Both individuals who choose to be accepted into the group as well as individuals who are born into it develop their identity under social pressure" (Castells 1997; Flint Ashery 2017).

The evidence from Boal (1976), Waterman and Kosmin (1988) suggests Jewish enclaves are present in the residential mosaic; however, there has been much debate concerning the actual existence of ethnic enclaves or ghettos in England (Johnston et al. 2007; Peach 1996). Peach (1996) creates a negative image through defining of ghettoes. Johnston et al. (2007) counteract this and suggest enclaves emerge where the host society is predominantly white, and the enclave is not considered as a cultural or political threat. Concluding from the literature, enclaves are present in British cities; however, it may be too extreme to classify them as ghettos—this term is predominantly used in American society where "hypersegregation" (Johnston et al.: 2007: 610) is witnessed. Territoriality, in this regards, is defined as a power strategy and means of exerting decisive and spatial influence in society (Gottman 1973; Soja 1989: 150), and the human territoriality is presented as defensive and aggressive behaviour (Altman 1975; Edney 1976; Malmberg 1980). Previous studies coined the term of "terrain of interests" between communities, which concerns social and/or interest groups competing over space and resources by using planning methods and citizens' rights discourse (Flint Ashery 2015, 2017). It is common to distinguish between primary, secondary and public territories according to the duration of occupancy, psychological centrality, degree and internationality of marking the behaviour, the degree of control and the response to invasion (Altman 1975; Altman and Chemers 1984; Paasi 1996). Both for the leadership and the community members, borders are essential for organising territoriality (Altvater 1998; Newman 2003).

Boal (1976) classic work on urban segregation patterns suggests that classification of spatial expressions of segregation can be considered in three types of grouping: colonies, enclaves and ghettos depending on internal and external forces of exclusion from the charter society. Ethnicity is often the primary cause of residential segregation, and this is evident from the literature, for Jewish communities in the UK and elsewhere. Boal (1976) states that there is a continuum between ghettos and enclaves. Ghettos occur where external forces dominant and enclaves occur where internal forces maintain exclusion. In reality, it is difficult to establish exclusion is forced or voluntary and identification is only possible by a consideration of the extent of assimilation. Giddens views both agent- and structure-centred approaches as fundamental to the analysis of borders (Giddens 1984: 17, 85). Whether gateways or barriers, areas of competition or cooperation, borders, according to Anderson and O'Dowd, "shape and are shaped by what they contain and what crosses or is prevented from passing them." They may be weak or strong, rigid in some cases, but flexible in others, "especially if the populations of a border area have similar languages and culture" (Kolossov 2005). Some are highly populated, others virtually empty (Anderson and O'Dowd 1999), display differences of inequality and asymmetry, whether economic, political, cultural or social.

6.2 The Litvish Community of Gateshead: Structuring Group Identity

The spatial division in the voluntary ghetto of Gateshead allows religiously based lives to remain cut off from external influences and safeguards the younger generation from the perceived threats of non-Litvish culture. This separation is a strategy which has been possible to accomplish due to the community's relative economic independence from the society at large and its occupation of a specific economic niche within the larger ethnic group, but it has also become desirable due to a particular sociocultural response to internal communal forces. Gateshead subgroup position is founded on the community's particular response to external and internal forces, and whose present ethnic communalities are expressed in ways which differ from other Jewish subgroups. Robinson et al. (2007) show that of the "1510 people recorded as Jewish in the Gateshead BNG area (a substantial undercounting of the population according to some local people), 50.1% were categorised as White British, 42.9% as White Other and 5.0% as other ethnic groups. It should also be acknowledged that many people refused to state their religion. More than half (60.2%) of the minority ethnic population in the Gateshead BNG area in 2001 were born outside the UK". It provides a subgroup position which is both recognised and pronounced by members as well as non-members of the community and could thus be specifically relevant for the residential patterns of tightly bonded religious groups and national/ideological minorities. The tangible outcome of these changes is that the community has not the only territoriality distinguishes itself within Anglo-Jewry but has also become increasingly isolated from the society at large.

Influenced predominantly by studies from the USA (e.g., Wirth 1927/1998) on the Jewish community's natural tendency to voluntary concentrate in particular areas of the city, the literature on British Jewry during the 1980s emphasised and "demonstrated the capacity for large numbers of Jews to remain within close geographical proximity to each other" (Newman 1985; Waterman 1989; Waterman and Kosmin 1988). Waterman and Kosmin (1988) suggest that Jewish communities form enclaves as residential shifts caused through socio-economic assent are mirrored by the voluntary change of cultural and religious institutions (Knox and Pinch 2010). A congregation, according to their view, and not segregation is a more accurate way of describing Jewish communities as internal cohesion is the dominant force preventing social assimilation. Valins (2003) argues that although the "drawing of boundaries by this community is defensive … Haredi Jews see these as key to securing their distinctive practices and beliefs. As a way of life, it is often highly segregated, exclusionary and is laced with a politics of who belongs and who is out of place. However, it is also based upon a philosophy of mutual support and aid, with a genuine spirit of community cooperation and social justice (if only for those considered to be 'same'). Blurring the borders may have the effects of reducing differences and inequalities between insiders and outsiders (Young 1986), but the result is that this also weakens internal ties."

Nonetheless, most Haredi communities are located in urban environments (Gutwirth 1996), and in line with Heilman (1973: 15) description, people in the community today conduct their lives with a minimum of involvement with the society at large. Moreover, as Valins notes, "the socio-spatial boundaries that members of the community (individually and collectively) draw are indeed not total or complete, and Haredi Jews 'cross the lines' in all sorts of ways, from their consumption of modern technological products, such as mobile phones, to the adoption by many of English as their first language." In this regard, the distinctive and unusual characteristics of Gateshead's Haredi Judaism provide a lens through which to consider how certain groups are seeking to resist the perceived "excesses" of (post) modernity, and thus how, to follow Valins' Broughton Parks' example, to stabilise and "secure their identities from outside 'corruption'. The construction of boundaries by this community reflects not just temporary or symbolic acts of defiance and resistance, but rather an all-encompassing, day-to-day attempt to police the lines. The very process of drawing these lines (consciously and unconsciously) defines and (re)creates their identities on a day-to-day basis by emphasising the imagined differences between themselves (as pure) and 'others' (as impure)—residential patterns of difference feed into idealised identities, which are positioned against the 'other' of those living beyond its spatial borders and outside of it". Despite the uniqueness and importance of this community, there has been little social research completed into the Jewish Community of Gateshead (Olsover 1980), research on the influence of internal processes within the Litvish community on residential aspects of Gateshead has been somewhat limited due to the lack of data.

6.3 The Jewish Settlement in Gateshead

The historical background of the Litvish community of Gateshead is linked with this of the nearby city of Newcastle (Fig. 6.1a, b). By the 1850s, the Jewish community of Newcastle was a well-established unit whose economic status had continued to improve. Though the majority of Newcastle's Jewish population were the members of the lower strata, the German immigrants that arrived during the eighteenth century had, with their German education and cultural knowledge, shown an unprecedented rate of social mobility between generations. By the end of the nineteenth century, an increasing number of Jews, mostly children of the German immigrants who had moved into wholesale and manufacturing thus making "the climb" into the middle class, had started to leave the west-end. They had moved into the more prosperous area of Jesmond. By the beginning of the twentieth century, about a hundred families (Olsover 1980) had settled in Jesmond creating the need for a synagogue in that area. Like other provincial communities, such as those of Leeds, Manchester, Hull, and many more, Newcastle hosted the development of an anglicised middle-class Jewish community who struggled for social acceptance and political freedom. Undeniably, the ensuing community did not represent a homogenous group regarding socio-economic conditions. There were on the one side the settled community, the shopkeepers, overseas importers and share brokers, the providers of essential goods and services to the English middle classes. On the other hand, there were the factory workers, tailors, slipper makers, glaziers and peddlers. Common to the majority of the community's members was the belief in the necessity to emphasise English ideals like the social sophistication, education and stability of the hosting English society. The inculcation of and adaptation to English ideas and aspirations had its price. It called for a slackening in religious observance and numerous religious-cultural compromises.

Copsey (2010) describes how in 1830, when "the Jewish community of Newcastle-upon-Tyne came together as a distinct body for the first time, only nine Jewish families resided in the city, and within eight years the city's first synagogue had been completed." Guttentag (1973) describes in his book "The Beginnings of the Newcastle Jewish Community", the top-down initiation that leads to the establishment of the Litvish community of Gateshead: "Dr Adler was statistically minded and was greatly concerned about education, believing that such information was valuable as the basis for further action. Questionnaires were sent to all congregations in the land, and the following replies came from Newcastle: "9 Baale Batim, 33 men and women 33 children. 1 Synagogue. Shochet Rev. Caro. Shammas I. Cohen. Mitzvos are sold. Public Bath Mikvah". Unlike London "which was inundated by shoals of immigrant poor following two or three ill-paid vocations", here there was an opportunity over a wide field to establish Newcastle as one of the leading "Jewish scholar communities".

(a)
(b)

Fig. 6.1 **a** Location of Gateshead; **b** a closer view of Gateshead

6.4 The Formation of Gateshead as "The Jewish University Town of Europe"

Throughout the next fifty years, the community steadily established itself with the number of Jewish families in the city rising to 150 (Olsover 1980: 17–28). By 1877, there were eighty children at the school. The Jewish Chronicle of 16 May 1879, indicated that "Jewish boys whom we should now call 'juvenile delinquents' were then gathered together by the authorities from various parts of the country there were never more than thirty at the Netherton Reformatory near Newcastle". In order to continue their Jewish training, and with the help also of the Newcastle Jewish community established in 1880 and lasted about thirty years: "Reverting to the day of the synagogue consecration, we note an interesting affirmation of current belief in the place of Anglo-Jewry in the Jewish world, at least of the West. To celebrate the day's event, a banquet and ball were held in the evening in the Queen's Head Hotel, Pilgrim Street. Dr Hermann Adler presided. The Rev. S. Harris, replying to the toast to the Chief Rabbi and clergy referred to the Chief Rabbi's influence in the development of the community in England. The present time, he said, was unequalled in the history of the community. The spread of education and other influences, perhaps, had been followed by increased materialism and scepticism, but the Jewish community in here had remained the most orthodox in the world".

Between 1880 and 1905, as Copsey (2010) further explains, "a new wave of migrants arrived, fleeing anti-Semitic persecution and economic hardships in Eastern Europe, and Newcastle's Jewish community experienced its most rapid growth. By 1900, the Jewish community in Newcastle numbered 1800" (see also Voeltz 1988). These new arrivals, for the most part, poor Jews from the Russian Pale of Settlement were pious and non-selective in their religious observances and posed an economic burden (Poll and Krausz 1975; Steinberg 1981). The Jews of Newcastle, like so many other communities in Britain, showed no pleasure at the arrival of the Eastern European immigrants (Neustatter 1955; Brotz 1955; Gartner 1974; Williams 1985), and did not want to be identified with the 'medieval attitudes' of the newcomers, neither in terms of class or ethnicity. As cultural differences quickly came to the fore, the resentment of the "old" immigrants towards their Eastern European brethren who were extremely 'visible' in their ethnicity, was due to the undesirable "image" which they presented to the non-Jews, and relations between the settled community and the newcomers grew somewhat fraught (Neustatter 1955; Brotz 1955; Gartner 1974; Williams 1985). To many Jews in Britain, by now a minority group whose aspiration it was to achieve social and political emancipation and to be middle-class and less concerned with outward signs of ethnicity, the arrival of the Yiddish speaking Jews with, their "foreign" manners their obvious poverty, distinctive dress and religious requirements represented an acute embarrassment. The newcomers were perceived as a threat not only to the community's internal organisation but also to its relationship with the general society. The established community typically regarded the East European settlers as their social inferiors, and it was feared that they would damage the acquired "good name" of the community.

The enrolment of the Eastern European Jews into the existing Newcastle Jewish community eventually exhibited within-group variations in religious practice (Goldstein 1974). "Some were unable to resolve their differences with the established community, and in the 1880s a small group cut loose, crossed the Tyne to Gateshead and formed a Haredi community. The founding of Gateshead's Academy for Jewish studies in 1929 provided a significant stimulus for Gateshead growth that soon earned an international reputation as "the Jewish university town of Europe" (Copsey 2010). The Gateshead settlement was directly founded on the importance of geographical isolation and the (re)creation of a shtetl-type life free from the immorality and profanity of the urban sphere. While the community in Newcastle has been gradually eroding, the Gateshead cluster has slowly expanded. Around 600 students passed through the academy between 1933 and 1939, considerably augmented the traditional community. The Jewish population in Gateshead in 1950 stood at 280 (Donbrow 1972), and by 1995, this figure had increased to 1100 (Willis 2004). The 2011 Census data show that of the "1510 people recorded as Jewish in the Gateshead BNG area (a substantial undercounting of the population according to some local people), 50.1% were categorised as White British, 42.9% as White Other and 5.0% as other ethnic groups. It should also be acknowledged that many people refused to state their religion."

6.5 The Litvish Community of Gateshead

The orthodox Jewish settlement of Gateshead is one of the most important seats of Jewish further and higher education, and therefore an attractive place to live for the Jewish Community, most of them are affiliated with the Litvish. The case study of Gateshead is an interesting case study to examine how high awareness of particularities of social identity and affiliation create underlying mechanisms used by members of the Litvish community of Gateshead to create an isolated and "contiguous" living space. The exclusive devotion to Torah education, together with the inherent conflict of religious lifestyle with modernity, has motivated voluntary territorial separation of the Haredi population into a ghetto of sorts characterised by clear boundaries and strict division from the outer world. These unique preferences provide an insight into the social, organisational and spatial mechanisms that form the group identity of the Litvish community and motivated individual residential choices in Gateshead.

The Jewish community in Gateshead occupies less than a square mile there and numbering several thousand in the Saltwell ward and Bensham area of Lobley Hill and Bensham ward (Fig. 6.2). This group is described by Willis (2004) as constituting "a minority within a minority". Its educational establishments, shops and businesses concentrate in an area loosely defined by Prince Consort Road, Saltwell Park, Whitehall Road, Bewick Road and Durham Road. This area is mostly a pre-First World War housing area, located in the north-eastern part of the Gateshead district and lies

Fig. 6.2 Litvish area of Gateshead

on the ridge sloping down towards the Team Valley, between the Durham road and the Team Valley Trading Estate.

6.6 Methodology: The Case Study of Gateshead

Although the Litvish community of Gateshead is not as closed as the Satmars, the community's members are dedicating themselves to Torah study, thus everything in their daily life which is not related to Torah studies—even strolling or talking about secular's subject—is prohibited. However, significant ties with people from the community were obtained through rabbis and key figures who had been contacted through previous studies in Jerusalem, connections through JFS school and even informal private relationships in Golders Green, all this help to increase the levels of trust and opened up the possibilities of women (mainly) and leaders to cooperate. In consultation with Rabbi Zimmerman, Grand Rabbi of the Gateshead Hebrew Congregation, and Rabbi Oppenheimer, a chief adviser to organisations and educational institutions in the Haredi community, it was decided to refrain from conducting an in-depth field survey in order not to bother the students and their families. However, familiarity with the community's social practices enabled me to receive the rabbis' support for conducting anonymous interviews in a manner that is not a systematic "census-like" survey.

Despite these limitations, this study is based on "real data", provided anonymously by the people themselves between March and June 2015. A mixed-method were used to analyse both primary and secondary data and to study the ways in which high awareness of particularities of social identity and affiliation create underlying mechanisms used by members of the Litvish community of Gateshead to create an isolated and "contiguous" Haredi living space.

Following Ethnographic "unstructured" methods (Becker and Geer 1970; Schatzman and Strauss 1973), semi-structured interviews (Mason 2005) were adopted as the main method of primary data collection. The interview was conducted in "naturalistic" fashion, often as a conversation (Clifford and Valentine 2003: 117), with the specific content and length of the interview being flexible and variable across the respondents to allow the interviewee to demonstrate detailed topic-related knowledge as well as covering more general themes. In addition to the Rabbis, 37 local females, from different parts of the Litvish enclave were interviewed, to clarify responses and the social processes that gave rise to spatial patterns (Scheurich and McKenzie 2008). Snowball sampling was in use, in which participants are recruited by chain referrals (Flint Ashery 2018a, b). Snowball sampling involves "asking respondents to nominate others, which are then approached by the researcher and asked to nominate others for the study. This procedure continues until names begin to reappear, or if the researcher feels, he/she has saturated the list of nominations" (see also Bryman 2009). Increasingly, studies are incorporating community members to help locate respondents, test survey instruments and interpreted results, with many studies also using community members or "insiders" to locate respondents to improve the overall quality of the data (Stepick and Stepick 1990).

Some secondary resources were also analysed including Gateshead council policies and strategies, planning applications, documents of the Housing Market Renewal Pathfinder, Gateshead Council publications for the public, Office of the Deputy Prime Minister and Home Office documentation and a variety of internet sources. Cross-referencing these resources enables an analysis of the socio-spatial boundaries that are used to define and mark out "friends" from "others" and to further contribute to broader geographic discussions about identity, segregation and religion.

6.7 Social Mechanisms

The individual identities of scholar Litvish community of Gateshead—more "Frum" compare to the Yeshivish community of Golders Green—are defined at all levels, classified by their rabbinical figures and Yeshivas (higher education institutes), which are all subject to hierarchical leadership. The Litvish communities' centres on the particular Yeshivas, which represents an important element in its members' socialization process. Both individuals who choose to be accepted into a specific Torah scholar's group and individuals who born into it develop their identity under social pressure. Chana (pseudonym, female, 40): "It's hard to be fully accepted into this dynamic community, and it's hard to last here, not only because of the level of Torah study but because the entire family must adhere to the norms of the community. But once you accept our norms, they encompass all aspects of your life, and each one of us is responsible one for the other and has the obligation to ensure that the other does not deviate from the norms (…) You are being watched all the time. For all our deeds, all the time, from choosing the names for our children to everything we utter, everything carries meaning. It is a great privilege to live here and raise our family, and therefore we all support each other, and the commitment is very high." Adults who live their lives in this Haredi Litvish community of Gateshead have made a conscious choice, a commitment, as to the manner in which they aim to live their lives. They have decided to live according to a specific code of conduct, based upon Talmudic ethics, which posits that concern with the well-being of one's fellow men is of greater importance than the fulfilment of their own selfish desires. Life in Gateshead is a severe and goal-directed undertaking in which it is both guided and constrained by in-depth knowledge of the Talmud. Since the Second World War, the formal requirement for Jewish education has been answered through the various Jewish schools, seminaries, Yeshivot and Kolelim that have been established worldwide. The informal learning, however, was provided by long-standing established orthodox families living together in communities saturated with the specific religious-cultural environment of Jewishness. Despite their Haredi way of life, members of this Litvish community found themselves involved with the society at large in numerous social activities, signalled an increasing willingness to adapt and adjust its lifestyle and sociocultural commonalities to those of the community at large.

During the last three decades, the surge back to orthodoxy has created a widespread demand for formal and informal Jewish education. The establishment of the various Jewish educational institutions in Gateshead meant that the community could provide and transfer the formal and informal Jewish knowledge. Today, Gateshead's involvement with the Jewish international scene in that its reputation as a centre for culture transmission attracts Jews from all over the world supplies the former with knowledge and competence within that community. Men and young boys from afar come to undertake higher religious studies, and this is one of the centres to where the world's Jewry turns to satisfy an increasing demand for an orthodox rabbinate. While some men come to the higher learning institutions of Gateshead to acquire a degree and consequent job prospects, to take up a position as rabbi to a community and others might only spend few years there before carrying on to universities or polytechnics, the dominant trend in Gateshead is to devote their lives to Torah study.

In addition to the intellectual effort, being part of the community also requires adherence to strict rules of behaviour following a particular world view and its emphasis on religious ethics. Batya (pseudonym, female, 34): "You know how it goes—'each of us is one small light, but together we shine bright' (A Chanuka song)? So if I donate today, pay a visit to the sick, give money to charity, or the synagogue, any action big or small can make a real difference. What I give now is an investment for our future—I know that if I'm in need, I will get support. That's why there's no homeless in our community, for instance. For our generation, it is most important to pass these values forward both from each one's perspective and for our community's interest. The ideals are identical, and the personal and communal interests should be the same." By generating strong separatist aspirations from the surrounding society and increasing sociocultural isolation, Gateshead established a position within the world of Jewry as a centre where one can gain, refine, expand and practise the increasingly valued "capital" of religious-cultural knowledge. The process of formalising culture transmission through great—levelled internationally recognised teaching institutions, in effect means that the community has created a Jewish International Torah capital which is produced, operated and transmitted within the limits of the community.

The Jewish schools mainly have pupils who are children of full-time students or staff of the higher education institutions or young men and women who have come from afar to study and experience Litvish way of Judaism (Mussar) in their lives. Community life revolves around the religious institutions, which together attract hundreds of students and families in the collective effort and desire to learn in the Litvish way of Torah environment. Students, those who in increasing numbers have families, can undertake their studies in Gateshead in institutions of the highest standards and can rest assured that their children are educated in the "right" schools. Nechamah (pseudonym, female, 46) describes the recent changes: "Our community has doubled since 2008, I think now there are at least 8000 families here. Some of them are families of rabbis who arrived here in recent years because at least seven Yeshivas opened here. Rabbi Zimmerman's commitment to education attracted a lot of girls from the age of 15. My daughters say that even from Israel they come here. It is known that a Litvish girl who wants someone who's well-developed in his learning a good match will look for a quality Shidduch here at Gateshead." Liphshiz (2017) cites Rabbi Schleider, a former leader of the Gateshead Jewish community and a historian: "In Lithuania, nobody wanted to marry a rabbi. It meant poverty to women who were not brought up inside the Haredi education system … But in Gateshead, this began to change in the 1940s, when a seminary for women and girls opened. That provided a nucleus of four hundred girls who want only someone with rabbinical status … this was a very powerful development that changed the face of the Gateshead Jewish community more than anything else."

The full-time learning lifestyle strengthens the Litvish status and the acceptance of its leadership by a broad swathe of the Haredi population, facilitating accumulation and maintenance of control over individuals' private and public life through social mechanisms. In this sense, residential decision-making to move to Gateshead mean the acceptance of the role of intellectual interest, philosophical direction and specific religious-cultural adherence over once life. The influence of the Rabbis in the upkeep of group cohesion, achieving and retaining the highest possible standards of ethics and moral, and in guiding people in the "right" direction, is widely recognised in the community. Esther (pseudonym, female, 40) describes the importance the community gives to the children being nurtured and educated according to the values of the community: "It is our duty to educate our children according to the Torah, and there is no compromise on this. The children are our investment. They are the future of our community. Until the age of mitzvot (12 for girls, 13 for boys), they are encouraged to behave as children. We will not send the children to schools where they will study subjects that are contrary to our values. This is unacceptable. We must protect them from foreign influences. So they (the local education authority) send reps to our schools trying to find where some of the children are. And they conditioned their support on our cooperation. But it's not that the children do not learn. They are well educated, all within the community." The official figures indicate that about five hundred of boys in Gateshead only are missing from the National Statistics' Primary school performance tables (Fig. 6.3), which shows the prevalence of schools in which religious studies were the focus of the core. Since the percentage of Haredi in Gateshead is about a tenth of their percentage in London, Fig. 6.4 indicates that the response of Gateshead's community to the Torah study discipline is very high.

The intellectual as well as human qualities of the rabbis and staff of the teaching institutions, the content and emphasis of their lectures, the devotion of the students who learn their combined with the community's specific stress on community's ethics, make for a particular interactive style. Along with early marriage and high fertility rates, the rate of employment is considered a direct cause of the high poverty levels of Haredi households in Gateshead (Housing Market Assessment 2017). In the demanding full-time study reality of Gateshead, where young and upwardly mobile (enhancer) families can be realising their preference to live within their birth community by being deserved—devoted to their community's lifestyle, the family's

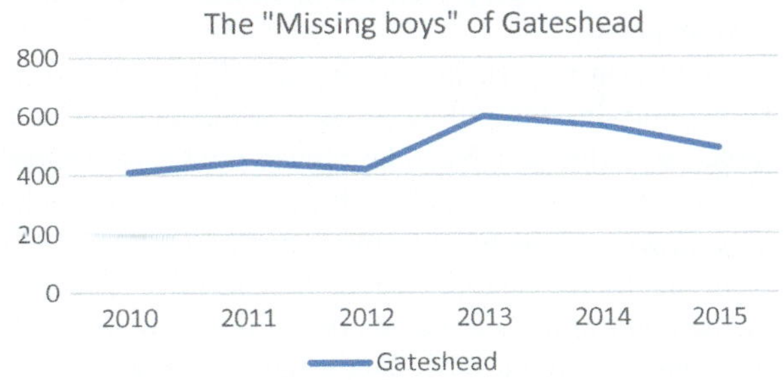

Fig. 6.3 "Missing boys" of Gateshead: the number of children who do not appear in the National Statistics' Primary school performance tables

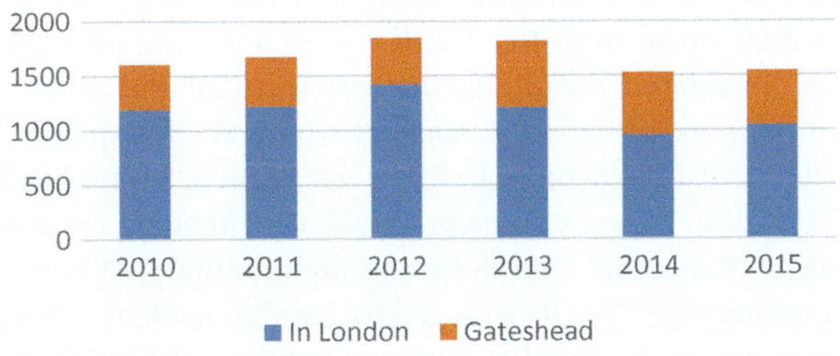

Fig. 6.4 Number of children who do not appear in the National Statistics' Primary school performance tables in London and Gateshead

economic dependence on the community is even higher, and proper group behaviour assumes greater importance as a means of receiving financial assistance.

In return for identification and affiliation with the group, the individual is expected to relinquish the perception of the free individual and to make his interests subservient to those of the group (Shapiro 2005). The needs of the individual are viewed as identical to the group interest; the individual's status assesses personal welfare within the group, and his contribution towards it and leavers are denounced. Chaya (pseudonym, female, 28) explains that the individual's status within the community has ramifications for his life course—from marriage prospects for his children to housing choices and options within the community: "We all have large families and we are all supported by the state. It is our Mitzvah to bring children, and the state is helping us to fulfil this mission by providing Child Benefit. It's about £20.70 per week for your first child and £13.70 a week for any further children. Most of my friends bring a child every two years because the body needs time to recover between births. Benefits help to meet housing and income needs thus allows our men to learn. Many of us are also supported by our parents, especially on special occasions that require a large expense. Because everyone wants to say that their son is Talmid Hacham, dedicated himself to the Torah. Because graduating from the Gateshead Yeshiva will improve not only one's knowledge of the Talmud but their chances of finding a good match for a wife, a quality Shidduch. It is also good for the younger brothers' matchmaking." Despite the efforts, both from inside and outside of the community have been made in recent years to boost Litvish participation in the civilian workforce; the level of involvement remains low. About 85% of Litvish men in Gateshead are devoted to full-time religious study, the rest work within the community, and most of the Litvish women are dedicating their life to educate their children according to the communal Litvish norms.

Homes in Gateshead are Spartan. Figure 6.5 shows a typical living room with heavy curtains that allow privacy, armchairs and sofas that open to beds, a study table and a dining table. The Jewish bookcase covers one of the walls of the room. Living within the surrounding local society, an area which is, at large, highly deprived, is regarded as beneficial to the Litvish community in their desire to remain isolated from the modern society. Rivke (pseudonym, female, 31) explains how the mission of the community permeates the individual's life: "This is not an instruction from the Torah, nor the Rabbis. But when you choose a particular lifestyle, it has a visual expression. There are decorative and discreet wallpapers, but usually, material objects are kept at a minimum. So most of the houses here are modest and anything that is a distraction from the Torah—even a picture of a landscape hanging on the wall—is considered unsuitable for the spirit of the place. I understand that this can be interpreted as poverty, and yes if you do not spend on the telly or decorations you can invest elsewhere, but it is our choice. It is a kind of austerity: when I buy new furniture, I want it to be practical, to withstand wear and tear for years and reflect no taste or income." Sarah (pseudonym, female, 18) describes her childhood within the community: "We fly to Israel for Simhas such as marriage and other events, and every two weeks do Shabbat with our families in Golders Green, but I grew up here, and I feel part of a broader mission of staying in Gateshead and raising children here, for

Fig. 6.5 A typical living room, which is also the kitchen and the dining space. On the other side of the room, there is a large cabinet of Torah books. All the single sofas are open to beds for the children returning from the boarding schools for Sabbath and for the married children who come with their families (*Credit* D. Flint Ashery, 2019)

our continuity. My cousins in London have board games, cards and piano which they play. Here it is less common. I had a great childhood here. I hardly had any games as a child, and when I grew up, I understood that we maintain a high Torah level means that we chose to avoid any 'unimportant' distractions." The contradiction with the surrounding local society serves as a means of highlighting their ethical values and decreases and minimised the risks of the young being attracted to the "other" world.

6.8 Organisational Mechanisms

Although Gateshead represents a high degree of consensus regarding religious precepts which form the basis for their sociocultural existence, it is essential to bear in mind that the Litvish community is a conglomerate of many nationalities and socio-economic backgrounds. Graham with Caputo (2015) argues that "Gateshead's position as a centre of orthodox Jewish study means that at least one-quarter of its Jewish population is students living in Yeshivas and seminaries." Gateshead accounts for almost a quarter (27%) of all Jews living in educational establishments in the UK. They farther explain that "Gateshead hosts several Haredi Torah study centres, and the vast majority of the students who attend these institutions come from outside Gateshead." The total number of students in the town more than doubled during the recent decade. According to JPR data (Staetsky and Boyd 2015), Gateshead has two religious seminaries for girls (aged 16–19) with about 600 students, and seven Yeshivas for boys (age 16–21) with about 1000 students, the Boarding School for boys, the Teacher Training College for Girls, the Primary Schools, the Jewish Girls High School and the Junior Yeshiva. Thus, more than 1600 orthodox Jewish students may study there. This number accords with a recent report on the community (Gateshead & Newcastle upon Tyne, Strategic Housing Market Assessment 2017) which states there is a permanent Jewish community of 3000 as well as 1500 Jewish students. However, the figures are not absolute: While the 2011 census recorded 2247 Jews living in households in Gateshead as well as 755 Jewish students in communal establishments, i.e. just over 3000 students in total, the local estimate that there are about eight thousand Litvish families living in Gateshead.

Rabbi Shraga Feivel Halevy Zimmerman is the Chief Rabbi of the Jewish community in Gateshead, one of the most significant figures in Gateshead's Litvish community. Rabbi Zimmerman took up his appointment and assumed leadership in 2008, in succession to Rabbi Bezalel Rakow, who died in 2003, and today he operates a "top-down" organisational mechanism that flows from the community leaders and heads of Yeshivas, via mashgichim (moral supervisors), to teachers and the students themselves. Roda (pseudonym, female, 27) describes that Rabbi Zimmerman runs the community as an executive director of a business: "Before the Rov the community was much smaller, there was barely a bakery and a small shop, there was no much for activities outside the synagogue. Most of my friends came from the city, and although we all want to live here, some of us found it difficult when they had children, and there were no suitable services for families. Sometimes you need professional advice. As the community grew and the Rov came from America, he managed to attract investments and services to our growing community. The families of my two friends of mine who have studied with me here in the seminary intend to return to raise their children here."

Shimon Guttentag, a senior director at the Yeshiva, says that the changes introduced by Rabbi Zimmerman from New Jersey have greatly improved community life. The rabbi tightened ties with the municipal authorities, which dealt with the needs of the community and with government officials, and led a massive investment in early childhood education. He also initiated the provision of social services, including counselling for couples suitable for the community, and it is possible to open a restaurant that operates at different times for men and women. Shulamit (pseudonym, female, 52) describes the changes that took place during the recent decade: "We left about twenty years ago when my husband got a job in London. But now the children grew up, and some of them moved here. So we sold the house in London, and now we live here together with the younger children and also can support our older ones who live here with their families. The community now is very different from what it used to be. But now my daughters say they will stay here after their graduation because there are services here that keep Yeshiva graduates who want to start a family." Graham with Caputo (2015) assumes that "the presence of large, but temporary Jewish student populations can have a somewhat distorting effect in some areas: The Jewish community grows by almost a half during term time due to students living in the city. Like tenure, accommodation type is mostly a product of location, life stage and wealth." Tzipporah (pseudonym, female, 52) explains: "There's a kind of dynamics here, people coming and going all the time. But on the other hand many returns after a few years when they have more children and have to cope with expensive living elsewhere. There is a cohesive group of local families and a smaller amount of mobile people, most of them are students. People say that the ratio remains the same, but the population grows."

The community maintain services and wield financial resources representing internal and external donations. Families with many children receive economic support through governmental child allowances, government assistance in housing, as well as specific funds by their community institutions. This support allows most Haredi men to continue their religious studies while the women generally take on the burden of socialisation of the young generation into the community with this one-way connection between the religious institution and in the absence of state oversight and religious monopoly, religious pluralism flourishes, with many rabbinical and various Yeshiva institutions. The combination of all these factors has transformed Gateshead from an old, sleepy community to a magnet for Haredi families. Explicitly, the majority of the Litvish living in the research area focus on residential relations when choosing a residence stated their preference to live among people of their group (Fig. 6.6). Esther (pseudonym, female, 23) explains why in Gateshead, where the Jewish population is young and urban, 83% of those who live in households lives in terraced houses: "It was clear to us that after the marriage we would like to live here because my husband had declared it before the wedding. And indeed the prices here are lower, and I met many new friends and there is a good framework for the children. We bought a terraced house here with a nice garden in a price for half a room in London. Everyone who lives on my street and the nearby streets is from us. I am quite certain that the prices here will rise because the community is expanding and everyone wants to live here. At least once a year we visit family in the United States or Israel so the proximity of Newcastle International Airport is very convenient."

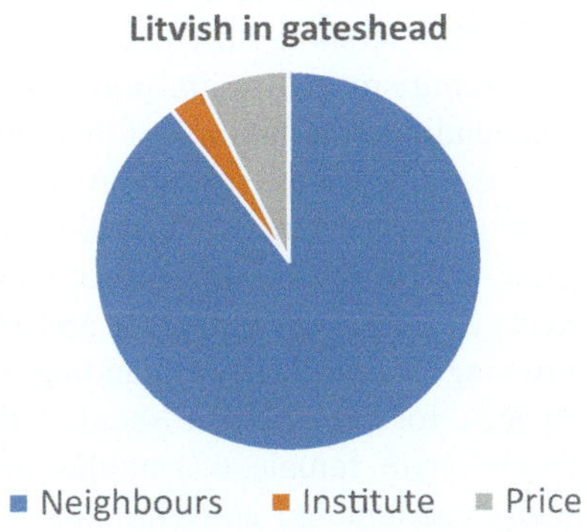

Fig. 6.6 Importance of apartment cost, neighbours' identity and proximity to institutions in apartment choice by Haredi groups, Stamford Hill (2015)

In addition to the opportunity to live a "pure" Litvish Torah lifestyle, Liphshiz (2017) explains that "Gateshead's appeal is an economic one: With housing becoming unaffordable in London, many religious Jews are looking northward for solutions. In recent years, some have moved to Manchester, the only other city in England that has seen its Jewish community grow". Luba (pseudonym, female, 38) chose Gateshead over Manchester for living in: "We also considered Manchester, but it is much more expensive and maybe a bit less (suitable) for our community. There is something about the fact that there are Yeshivas and that Gateshead is isolated from everything that makes life so comfortable to concentrate on what matter. It was essential for us to live within the community. I work in a kindergarten in the synagogue, and my husband is Yeshiver. We bought a semi-detached house within a ten-minute walk from the synagogue, and when my parents visit us, my father always says it was his dream to live so close to the synagogue. This internal division and articulated segregated attitudes hint at the powerful social structure and active social organisation operating in this group, in the form of tight relations between the community members and the leadership.

Aside from being a study institute, the Yeshiva provides systems of services that cater to the educational and social welfare needs of the individual and the family. Through the institutions, the leadership regulates a system of control and supervision following the desired codes of behaviour. Devoted representatives of the leadership dictating the substance of such concepts as "a good family", "proper behaviour", "Tsaddik" (a saintly, righteous individual) and structuring commitment and obedience (Heilman and Friedman 2010). By this mediation, the way in which individuals perceive themselves, their families, and their community is reinforced and perpetuated. The parents' generation adopts the lifestyle dictated by the leadership and passes on its values to the younger generation. One of the most salient boundary markers in Gateshead between the Litvish community and the society at large (as well as other Jewish subgroups), which is not objectively apparent to the outsider, is the degree of preoccupation with ethical and moral conduct prevailing in the community. By introducing and providing interpersonal ethics with a central and pervasive role in people's lives, people have had to seek answers as to how this type of conduct is secured. The latter involves an understanding of the locus for the origin of the individual's particular course of action and the means for governing this operation.

Rabbi Gershon Miller, a senior educator at the Yeshiva, refers to the rise in anti-Semitism in Britain, noting that the community is in good relations with the surrounding population and that although the community has grown, few incidents have occurred. Ruth (pseudonym, female, 24) argues that the community has learned from past experiences: "I have heard that in other places, as the Haredi population grew, our different appearance outside the community's area drew attention and many expressions of hostility and antisemitism aroused. The community here is still growing, but gradually expanding outward. I think we have learned from past experiences, for example, in Shenley, that even recently there were incidents. Tziviah (pseudonym, female, 67) another member of the Litvish community says: I have not really encountered antisemitism, it is more of people staring at us because we do look different. People probably expect us to want to talk to them. But that's not our way." Rabbi Schleider says that the Gateshead Yeshiva doesn't have any guards or even a perimeter fence: "Gateshead is also one of the few places in Britain with a large Jewish community where anti-Semitism is not a primary concern … We are mindful of the situation today but thankfully, we've not had a serious incident and it's not a major problem affecting us" (Liphshiz 2017).

6.9 Spatial Mechanisms

6.9.1 The Effect of Group Behaviour on Residential Conditions

As housing markets across the north-east of England are undergoing significant restructuring, housing and neighbourhood declines are not uniform across Gateshead. Leather et al. (2002) suggest that particular neighbourhoods within Gateshead need to adopt planned approaches to diversify tenure and increase owner occupation. In conjunction with national urban and social renewal policies, Gateshead has begun to restructure and review local housing renewal policy to deal with the changing needs of its communities. According to Gateshead Council (2017), housing is an essential issue for the Jewish community. Jewish family households tend to be larger than average, and the internal requirements can be challenging to achieve in an area of terraced houses and flats with small garden spaces and backyards (Housing Market Assessment 2017).

Most of the Litvish Yeshivas can help community members by offering loans on good terms for renting apartments owned by the Yeshiva or purchasing new apartments within the delineated areas that are approved by the leadership. Gateshead & Newcastle upon Tyne, Strategic Housing Market Assessment (2017) found that "The average household size from survey respondents was 5.9 persons per household, which is far higher than the average household size across Gateshead as a whole. The majority of respondents presently reside in terraced homes (60%) or flats (23.6%). A large proportion of respondents are owner-occupiers (54.5%) or in private rental (37.3%) but very few reside in council (1.8%) or housing association (3.6%) homes. Nearly, 40% of respondents consider their home not to be suitable for their household's needs, and nearly 20% of respondents consider the condition of their present home to be inadequate. Many are either currently overcrowded or anticipate being so in future. Over 37% of respondents are expecting to move home within Gateshead within five years. Over 50% of respondents envisage needing more rooms in the next 5–10 years. Less than 12% expect to downsize within that period". The report indicates that the cultural needs of the community require the housing provision to be convenient for the synagogue and educational establishments. Many in the Jewish Community aspire to owner occupation inwards because of the sense of community and the proximity to cultural facilities such as specific schools and Kosher shops, putting pressures upon the housing stock. However, typically lives in

Table 6.1 Proportion of overcrowded households 2011 and change 2001–11 by tenure

Gateshead	Occupancy rating (rooms)						Occupancy rating (bedrooms) 2011	
	2001		2011		Net change 2001–11			
	N	%	*N*	%	*N*	%	*N*	%
Owned	1591	3.2	1386	2.7	−205	−18	1021	2.0
Private rented	545	8.5	1381	11.6	+836	+35	707	5.9
Social rented	3009	10.4	2721	10.9	−288	+5	1375	5.5
All households	5145	6.1	5488	6.2	+343	+1	3103	3.5

Note Overcrowded households are considered to have an occupancy rating of −1 or less
Source UK census of population 2001 and 2011

large households, there is a shortage of suitable properties for them to purchase. This process weakens Bensham's economic base, harms the level of services to residents and increases the level of overcrowding and the level of segregation.

According to 'Strategic Housing Market Assessment' report (2017), the level of density in Bensham is of 23% while the level of density in Gateshead is 14%. Bensham is within the highest 30% of deprived areas in England as defined in the national Index of Multiple Deprivation. Summarising the data from Gateshead report (Strategic Housing Market Assessment 2017), Table 6.1 shows that for Gateshead, "overcrowding increased from 5145 to 5488 households (an increase of 343) over the 10-year period 2001–11. This represents a percentage growth of 1%, which is much lower than comparator areas; Leeds (17%), Liverpool and Knowsley (18%), Manchester and Salford (52%) and Sheffield (50%). It is also lower than the national increase for England (23%). When considered by tenure, overcrowding has reduced by 205 households in the owner-occupied sector and reduced by 288 households in the social rented sector; however, the number of households overcrowded in the private rented sector has increased from 545 to 1381, a growth of 836 households over the 10-year period. The percentage of overcrowded households in the private rented sector has also increased from 8.5 to 11.6% (a percentage increase of 35%). When considered by tenure, overcrowding has reduced by 159 households in the owner-occupied sector and reduced by 165 households in the social rented sector; however, the number of households overcrowded in the private rented sector has increased from 1962 to 4303, a growth of 2341 households over the 10-year period. The percentage of overcrowded households in the private rented sector has also increased from 13.3 to 18.0% (a percentage increase of 35%)." Statistical analysis showed that density ranged from some streets with no Jewish settlement at all to others with 95% Jewish inhabitants. Following Vaughan and Penn research (2006), the absence of Jewish settlement in particular streets in such situations can be attributed to prejudice but is also attributable to the desire of the Jewish immigrants to live with others from the same religious background. An analysis of the relationship between the density of Jewish inhabitants and local integration shows a strong reverse correspondence between density and spatial measures. This suggests that the higher the density of Jewish settlement, the lower the spatial integration of the street in which they live.

6.9.2 The Formation of Territorial Boundaries

The shared interests and common behaviour intensify influence and social exchange and encourage cooperation within the delineated enclaves (Heilman and Friedman 2010). A range of Yeshivas operating under the Litvish umbrella competes among themselves to intensify their religious specialisation and shore up their social prestige. Management of the religious economy encourages development with the aim of growing the community and competing over influence and attracting resources. Apart from differences in dress, religious customs and values, there are other elements which underpin the ideology of close social association (Cohen 1982). One could say that Gateshead constitutes a kind of "moral community" (Bailey 1971; Larsen 1982; Eriksen 1993), where people are aware that their particular lifestyle distinguishes them from others.

The literature on the historical context of the Gateshead Jews suggests that the community segregation can too be linked to enclaves mirroring patterns of other communities within the country (Olsover 1980). Jewish communities provide a classic example of where segregation is promoted "to preserve and promote their cultural heritage" (Boal 1976: 48). Within Jewish enclaves, the rules of the Sabbath, preparation of food, dietary requirements, synagogue attendance and community duties maintain the enclave segregation especially within Haredi communities like Gateshead. The Jewish community of Gateshead had maintained segregation through being a geographically, close-knit community which has developed its cultural practices and distinctive way of life without any concern for the opinion of the general public who live around their created enclave. The boundaries (Fig. 6.7) which separate the Jewish community from its surroundings are well-defined. Moreover, in recent years, along with the increase in the community population, new institutions have been opened some distance from the current enclave towards the northeast. New properties were quickly populated in the area between the institutions and the already established enclave indicates that the institutions are acting as a spearhead for expanding the enclave. The border between the Haredi enclave and its surroundings is also accompanied by price line (Fig. 6.8). The prices within the enclave preserve a fixed gap during the years and are higher by 20–25% than the general prices of properties in Gateshead.

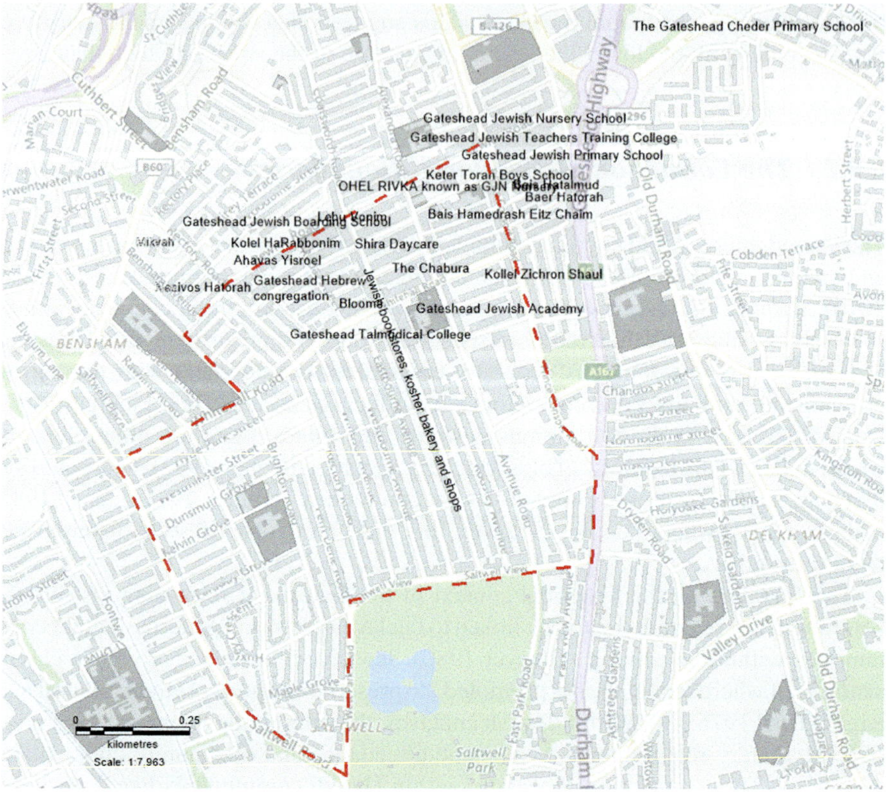

Fig. 6.7 Boundaries which separate the community from its surroundings are well-defined

Fig. 6.8 Internal housing market in Bensham (Zoopla 2019)

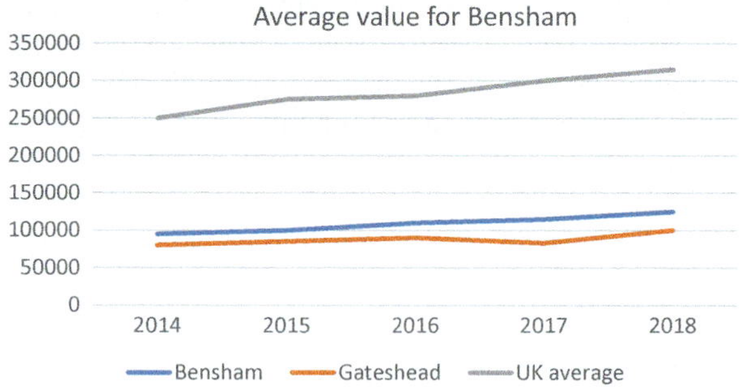

6.10 Discussion: Torah Position as an Engine to Acquire Spatial Homogeneity

The renewed interest and deep concern with Torah Lishmo and the specific interest in Mussar have created a community whose primary concern is that of living lives which are in meticulous compliance with Torah precepts and with particular attention given to interpersonal ethics. Although there are other Haredi communities in Britain, Gateshead's specific role within the world of Jewry and its relative independence from the society at large's economic structure provides the Litvish community there with a particular position. It is an acquired position which is based upon a combination of their reputation as one of the world's most important centres of formal Jewish culture transmission and the homogeneity and level which the community demonstrates regarding their deep-felt religiosity and devotion to Jewish ethics. It would seem that Gateshead has in some ways become a yardstick with which members of other Jewish communities might evaluate their levels of adherence to, as well as a choice of sociocultural commonalities upon which they base their ethnic identity. Throughout fifty years, these changes had restyled interpersonal interactions and brought about a renegotiation of particular sociocultural constructs.

The study of Gateshead provides a useful link between discussions of identity and place and more traditional geographies of segregation and territoriality. The spatial division in the voluntary ghetto of Gateshead allows religiously based lives to remain cut off from external influences and safeguards the younger generation from the perceived threats of non-Litvish culture. This behaviour is a strategy which has been possible to accomplish due to the community's occupation of a specific economic niche within the larger ethnic group, but it has also become desirable due to a particular sociocultural response to internal communal forces. Gateshead subgroup position is founded on the community's specific response to external and internal forces, and whose present ethnic communalities are expressed in ways which differ from other Jewish subgroups. Gateshead's Jewish population is growing expeditiously, provides a subgroup position which is both recognised and pronounced by members as well as non-members of the community and could thus be specifically relevant for the residential patterns of tightly bonded religious groups and national/ideological minorities. The tangible outcome of these changes is that the community has not only distinguished itself within Anglo-Jewry but has also become increasingly isolated from the society at large. People in the community today conduct their lives with a minimum of involvement with the society at large as Heilman (1973) describes: those who are "too modern" are criticised for the insufficient Frumkeit, the "too frum" is faulted for the opposite excess (ibid; 15). The case study of Gateshead, in this regards, indicates that belonging to the strictly—religious community is the primary source of strength and vitality—of minimising the risk to the self of interacting with others, in Giddens' terms—and the stronger it gets, the more impact it has on the individual's norms and lifestyle.

Gateshead and Newcastle upon Tyne Strategic Housing Market (2017)/report of findings state that nationally, faith community awareness is significantly increased but what is a more important consideration is how this policy is developed to the micro-level. Subregional and private sector involvement has been a key to the success of current renewal programmes alongside successful mechanisms of Jewish participation. Although Jewish-led home improvement is beginning to occur, the Jewish community internally have little community-wide policy or provision for grassroots housing renewal. An emerging concern is the lack of internal Jewish system as the community already has a well-established social infrastructure, and from the municipality point of view, there is a little reason to suggest why this has not been extended to housing as a fundamental policy progression. Housing has been established as a critical issue; however, it has not been placed at the top of the community's agenda. According to Gateshead municipality, policy intervention is a fundamental first step in recognising and addressing this specific housing need, however, what is crucial is the enactment of these policies within the community.

The Litvish community congregates within a defined area, where dedicated services and institutions are located and where they can preserve their cultural heritage and identity within a designated living space. This area is surrounded by a clear price line, which defines and gradually increases by the desire of the individuals to buy or rent properties in this area. With the growth of the population and the growing pressure on the enclave, official allocations are inadequate and are the community turns to unofficial modes of group action to achieve their goals. Additional community institutions were established in an area located north-eastern to the enclave to occupy space and claim resources. Residential buildings are gradually purchased in the area that is located between the established enclave and the new institutions to expand the enclave. The liberal system is thus unable to regulate resources in terrain where sectoral interests clash. It can be therefore assumed that with the enclave slipping towards the northeast, prices will rise in the region, respectively.

References

Altman I (1975) The environment and social behavior: privacy, personal space, territory, and crowding

Altman I, Chemers MM (1984) Culture and environment (No. 2). CUP Archive

Altvater E (1998) Theoretical deliberations on time and space in post-socialist transformation. Reg Stud 32(7):591–605

Anderson L, O'Dowd J (1999) Borders, border regions and territoriality: contradictory meanings, changing significance. Reg Stud 33(7):593–604

Bailey FG (1971) Gifts and poison, the politics of reputation. Basil Blackwell, Oxford

Bankston CLIII, Zhou M (1995) Religious participation, ethnic identification, and adaptation of Vietnamese adolescents in an immigrant community. The Sociological Quarterly 36:3

Becker HS, Geer B (1970) Participant observation and interviewing: acomparison. In: Qualitative methodology: firsthand involvement with the social world, pp 133–142

Boal FW (1976) Ethnic residential segregation. Soc Areas Cities 1:41–79

Brotz HM (1955) The outline of Jewish society in London. In: Freedman M (ed) A minority in Britain. Valentine, Mitchelle and co, London

Bryman A (2009) Social research methods. Oxford University Press, Oxford

Castells M (1997) The power of identity. Volume II of the information age: economy, society and culture. Blackwell Publishers, Malden, MA, Oxford

Clifford NJ, Valentine G (2003) Key methods in geography. Sage, London

Cohen AP (1982) Belonging: the experience of culture. In: Cohen AP (ed) Belonging: identity and social organization in British rural cultures. Manchester University Press

Copsey N (2010) The English defence league: challenging our country and our values of social inclusion, fairness and equality.

Donbrow M (1972) They docked at Newcastle and wound up in Gateshead. Rubin Mass, Jerusalem

Edney JJ (1976) Human territories: "Comment on functional properties". Environ Behav 8(1):31

Eriksen EH (1993) Reproducing moral agents: the concept of human nature in an orthodox Jewish community. Durham E-theses

Flint Ashery S (2015) Public welfare or sectarianism: a new challenge for planning. Plan Theory & Pract 16(3):299–318

Flint Ashery S (2017) 'Decision not to decide': a new challenge for planning. Eur Plan Stud 25(6):1076–1098

Flint Ashery S (2018a) Schelling-type micro-segregation in a Hasidic enclave of Stamford-Hill. Hous Stud 33(7):1038–1059

Flint Ashery S (2018b) Micro-residential dynamics. A case study of Whitechapel, London. Springer. ISBN 978-3-030-00601-3

Flint Ashery S, Benenson I, Alfasi N (2014) Group and collective behaviour: residential patterns in Kiryat HaYovel, Megamot–Szold Institute for Behavioural Studies (Hebrew)

Gartner LP (1974/2001) The Jewish immigrant in England, 1870–1914. Mitchell Valentine & Company

Giddens A (1984) The constitution of society: outline of the theory of structuration. University of California Press, Berkeley and Los Angeles

Goldstein S (1974) Jewish religiosity. In: Skiare M (ed) The Jew in American society. Bihman House Inc, New York

Gottman J (1973) The significance of territory. University of Virginia Press, Charlottesville

Graham D, Caputo ML (2015) Jewish familes and Jewish households: census insights about how we live

Guttentag GD (1973) The beginnings of the Newcastle Jewish community. Trans Misc (Jew Hist Soc Eng) 25:1–24

Gutwirth J (1996) Hasidism and urban life. Jew J Sociol 32:105–113

Heilman SC (1973) Synagogue life: a study of symbolic interaction. The University of Chicago Press, Chicago

Heilman S, Friedman M (2010) The Rebbe: the life and afterlife of Menachem Mendel Schneerson. Princeton University Press

Johnston R, Poulsen M, Forrest J (2007) The geography of ethnic residential segregation: a comparative study of five countries. Ann Assoc Am Geogr 97(4):713–738

Knox P, Pinch S (2010) Urban social geography: an introduction, 6th edn. Pearson, London

Kolossov V (2005) Border studies: changing perspectives and theoretical approaches. Geopolitics 10(3):606–632

Komarova M (2008) Shared space in Belfast and the limits of a shared future. In: Conflict in cities working papers series 3, Conflict in cities and the contested state: everyday life and the possibilities for transformation in Belfast, Jerusalem and other divided cities. UK Economic and Social Research Council Large Grants Scheme, RES-060-25-0015, 2007–2012

Larsen SS (1982) The two sides of the house: identity and social organization in Kilbroney, Northern Ireland. In AP Cohen (ed) Belonging: identity and social organization in British rural cultures. Manchester University Press, Manchester

Leather P, Lee P, Marie A (2002) North East England: changing housing markets and urban regeneration. The University of Birmingham, Birmingham

Liphshiz C (2017) https://www.jta.org/2017/10/03/global/why-orthodox-jews-are-flocking-to-this-gritty-english-town, 3 Oct 2017. Jewish Telegraph Agency

Malmberg M (1980) Human territoriality: survey of behavioural territories in man with preliminary analysis and discussion of meaning. Mouton Publishers
Mason P (2005) Visual data in applied qualitative research: lessons from experience. Qual Res 5(3):325–346
Neustatter H (1955) Demographic and other statistical aspects of Anglo-Jewry. In: A minority in Britain: social studies of the Anglo-Jewish community, vol 59
Newman D (1985) Integration and ethnic spatial concentration: the changing distribution of the Anglo-Jewish community. Trans Inst Br Geogr 360–376
Newman D (2003) On borders and power: a theoretical framework. J Borderl Stud 18:13–24
Newman D (2006) The lines that continue to separate us: borders in ourborderless' world. Prog Hum Geogr 30(2):143–161
Olsover L (1980) The Jewish communities of North East England 1755–1980. Ashley Marks publications Co, Newcastle Upon Tyne
Paasi A (1996) Territories, boundaries and consciousness. Wiley, New York
Peach C (1996) Does Britain have ghettos? Trans Inst Br Geogr 216–235
Poll S, Krausz E (1975) On ethnic and religious diversity in Israel. Institute of the Studies of Ethnic and Religious Groups, Israel
Pratt G (1999) Geographies of identity and difference: marking boundaries. In: Massey D, Allen J
Robinson D, Reeve K, Casey R, Gousie R (2007) Minority ethnic residential experiences and requirements in the bridging Newcastle Gateshead area
Schatzman L, Strauss AL (1973) Field research: strategies for a natural sociology. Prentice Hall
Scheurich J, McKenzie K (2008) Foucault's methodologies. Collect Interpret Qual Mater 3:313–349
Shapiro M (2005) How did fundamentalism manage to infiltrate contemporary orthodoxy: response to Heilman. Contemp Jew 25(1):273–278
Soja EW (1989) Postmodern geographies: the reassertion of space in critical social theory. Verso
Staetsky D, Boyd J (2015) Haredi rising: what the demography of British Jews tells us about the future of the community. Institute for Jewish Policy Research (JPR)
Steinberg S (1981) The ethnic myth. Athenuni, New York
Stepick A, Stepick CD (1990) People in the shadows: survey research among Haitians in Miami. Hum Organ 49(1):64–77
Strategic Housing Market Assessment (2017) Gateshead & Newcastle upon Tyne, opinion research services, report of findings, August 2017
Valins O (2003) Defending identities or segregating communities? Faith-based schooling and the UK Jewish community. Geoforum 34(2):235–247
Vaughan L, Penn A (2006) Jewish immigrant settlement patterns in Manchester and Leeds 1881. Urban Stud 43(3):653–671
Voeltz RA (1988) '… A good Jew and a good Englishman': the Jewish Lads' Brigade, 1894–1922. J Contemp Hist 23(1):119–127
Waterman S (1989) Jews in an outer London borough, Barnet. Department of Geography, Queen Mary College
Waterman S, Kosmin BA (1988) Residential patterns and processes: a study of Jews in three London boroughs. Trans Inst Br Geogr 79–95
Williams B (1985) The making of Manchester Jewry. Manchester University Press, Manchester
Willis B (2004) Housing market renewal: a regeneration and renewal special. Regeneration 30 Jan 2004, London
Wirth L (1927/1998) The ghetto. Am J Sociol 33(1):57–71
Young IM (1986) The ideal of community and the politics of difference. Soc Theory Pract 12(1):1–26

Chapter 7
The Litvish Communities of Golders Green and Gateshead

Abstract This chapter compares the mechanisms used by members of the Litvish communities of Golders Green and Gateshead to construct a private and public "demarcation terrain" between the abstract obligations and the spatial reality in which they are applied. In Golders Green, the Litvish community organises itself while competing within a diverse population. This reality brings exposure to and rationalisation of the employment market and consumer culture, with a weakening of religious reliance. Personal responsibility allows the Litvish man a degree of flexibility when combining religious studies and employment, and relative freedom in adapting the communal lifestyle to his abilities and professional preferences. In Gateshead, the community is organised in a "top-down" manner and arranged around its bold institutions. Social norms and mechanisms of supervision create a high level of awareness of the leadership's values and expectations. In Gateshead's municipal politics, the awareness to the Litvish Haredi group increases, as they are highly aware of the demographic characteristic and the unique housing needs of this community. These preferences and tendencies find expression in the residential enclaves: the clustering within a community enclave, the extremism against other communities in Gateshead, and the high level of residential heterogeneity within other Jewish and non-Jewish communities in Golders Green, all award the delineated area its spatial boundaries.

Keywords Group behaviour · Cultural dominance · Territoriality · Voluntary ghetto · Residential relations · Residential enclaves

These first two chapters compare the mechanisms used by members of the Litvish communities of Golders Green and Gateshead to construct a private and public "demarcation terrain" between the abstract obligations and the spatial reality in which they are applied. The existing literature assumes that group identity and intergroup competition in the modern city can affect religious development. This is especially true for the Litvish group, which operates social, organisational and spatial mechanisms to form delineation between religious obligations and the reality, to ensure the survival and resilience of the communities in competition with neighbouring groups.

In Golders Green, the Litvish community organises itself while competing within a diverse population. Individuals and families themselves bear the responsibility for building and maintaining the next generation's communal identity. This reality brings

S. Flint Ashery, *Spatial Behavior in Haredi Jewish Communities in Great Britain*, The Urban Book Series, https://doi.org/10.1007/978-3-030-25858-0_7

exposure to and rationalisation of the employment market and consumer culture, with a weakening of religious reliance. Personal responsibility allows the Litvish man a degree of flexibility when combining religious studies and employment and relative freedom in adapting the communal lifestyle to his abilities and professional preferences. Elements of socialisation and choice in the life cycle—the individual's attachment to the community from infancy, his religious education, the supervision of his young family and his position as an independent adult who makes a living in order to provide an economic basis for the life cycle to continue—strengthen the individual's identity on an ongoing basis through comparison with the external society.

In Gateshead, the community is organised in a "top-down" manner and arranged around its bold institutions. The individual's dependence on the ongoing support of the community's institutions creates a high level of awareness of the leadership's values and expectations. Social norms and mechanisms of supervision attach the men to the Yeshiva world and prevent most of them from joining the workforce. Dependence of a broad population of full-time scholars and their families on shrinking outside support raises the spectre of economic collapse, while competition over the allocation of resources creates internal conflicts. In Gateshead's municipal politics, the awareness to the Litvish Haredi group increases, as they are highly aware of the demographic characteristic and the unique housing needs of this community. As an example, designated construction within the segregated residential areas is evidence of the growing awareness of the municipal to the group's needs.

These preferences and tendencies find expression in the residential enclaves: the clustering within a community enclave, the extremism against other communities in Gateshead and the high level of residential heterogeneity within other Jewish and non-Jewish communities in Golders Green, all award the delineated area its spatial boundaries. In Golders Green, the constant encounter with other, mainly non-Jewish groups, serves to bond the community and soften its socio-organisational-spatial delineation. It is precisely the reality of ongoing interaction, entailing constant awareness and ongoing consolidation of identity concerning others and modernity that has the power to structure a community with a robust and positive group identity. This relative flexibility allows members of the community to adopt a more materially comfortable lifestyle while maintaining communal norms while living alongside different populations. It is possible that it is precisely these individualist modes of integration that contribute to the community's survival. In contrast, the view that a Haredi person is more likely to adopt a different Jewish identity than to abandon his religion altogether has prompted the Litvish leadership in Gateshead to take rigid boundaries that have led to the creation of homogeneous residential patterns.

The comparison between the inner Litvish communities of Golders Green and Gateshead raises the possibility that countering processes, which make religion a choice, there are opposite trends which strengthen the Litvish identity. Adopting some flexibility of the mechanisms involved in maintaining Litvish communal identity in Golders Green might enhance the resilience of the Litvish community of Gateshead. These mechanisms could allow for a combination of work and study and prevent abandonment of communal identity when the government's support is limited since the collective identity of those remaining within the Yeshiva framework solely because of the economic support that it offered would likely be weakened. In adopting that flexibility, attention should be paid to the ramifications of transferring responsibility for the socialisation from the institutions to the family, since a considerable number of women earners may turn back to the home to take a significant role in this process. The men, whose level of general education is lower than that of the women, would require professional training. Such training, at the accepted standard for the UK's workplace, could increase income levels for all Haredi earners, and Litvish earners in particular, to the national average. Haredi involvement in the workforce would create day-to-day contact among different sectors. Greater equality in bearing the economic burden and the possible contribution of Litvish workers to the UK's economy could increase tolerance among the general public. From the Litvish perspective, adopting this attitude can decrease the ignorance towards their sector.

Despite ever-increasing global interconnections and processes of spatial integration, as we saw in Golders green, the example of the Litvish community Gateshead demonstrates how patterns of hyper-segregation remain and expand as a contemporary urban phenomenon. Litvish society shares the view that living in Gateshead—an essential centre of Torah Judaism outside Israel—demands a higher level of commitment to religious obligations. The Litvish community of Golders Green desires to achieve full integration within the socio-economic and political life of the country, but also wish to maintain his/her traditional observance. This section of the community has tended to move out of the inner city to the suburbs but in one specific direction—to Gateshead, thus maintaining community life and also providing the necessary threshold for the establishment of new community institutions and services. In Gateshead, Litvish society has tended to separate and insulate itself from the wider community (Kosmin and Waterman 1989). It closes itself off from the surrounding non-Haredi Jewish society and cuts off the younger generation from employment frameworks and exposure to modern, professional education. Hence, while in Gateshead the Litvish enclave is designated, the Golders Green community is more integrated with the general society. In both areas, the local authorities provide the necessary regulations for maintaining Haredi public space. The local authorities are setting Eruv wires (ritual enclosures making an area a "private domain", which makes it permissible to carry objects on the Sabbath) and allocating buildings for religious services for the group.

In the context of an unstable world economy and social changes, the Litvish communal identity faces daily challenges. This may have particular relevance to the question of processes of integration or rejection of communities that refrain from participation in the workforce. Is it possible that religious or ethnic population groups avoid integrating into the workforce owing to fear (on the part of the leadership or the members themselves) of a weakening of communal identity? Could the bond between members of such groups and the general society be strengthened through reinforcement of their collective identity, such that they could be integrated without the threat of disintegration? Could a strengthening of group resilience and individual identity lead to decreased hostility among different groups? The present study compares the mechanisms serving the Litvish communities in Golders Green and Gateshead in structuring the individual and collective delineation between halakhic requirements and the reality within which they are implemented and exposes value-based elements that ensure the community's survival and resilience. The case study shows how the power of religious identities has not waned with the rise of modernity. How many people's lives still revolve around matters of faith has significant policy implications (Ouseley 2001).

The Haredi community in Gateshead shows how a way of life with clear socio-spatial boundaries and a stubborn resistance to many forms of change is undergoing significant demographic growth and is thriving. Analysing the patterns and processes that underpin the practices of this community thus ties into growing calls for geographers to recognise the power of religion fully to shape the identities of the vast majority of the world's population (Holloway and Valins 2002). Religion can no longer just be ignored or conflated with conceptions of race, class or gender when analysing the geographical worlds that humans make and inhabit (Kong 2001). Identification of a flexible and adaptive approach to delineation setting in the spatial reality of these two neighbourhoods broadens our knowledge about the spatial behaviour of the Haredi population, and the ramifications of changing policy on the strengthening of individual and collective identity given the challenge of exposure to "others". The respective delineations exposed in the two communities raise the possibility of their adoption and implementation in other enclaves. The comparison between these two Litvish communities indicates that social, organisational and spatial mechanisms existing within Haredi society may exist within other communities as well and may have the power to influence society at large.

References

Holloway J, Valins O (2002) Placing religion and spirituality in geography

Kong L (2001) Mapping 'new' geographies of religion: politics and poetics in modernity. Prog Hum Geogr 25(2):211–233

Kosmin BA, Waterman S (1989) The use and misuse of ethnic names in Jewish social research. Jew Popul Stud (19):1–9

Ouseley H (2001) Community pride not prejudice: making diversity work in Bradford. Bradford Vision, Bradford, p 11

Chapter 8
The Hasidic World

Abstract Hasidic Judaism, meaning "piety", is a prominent sector, within strictly—orthodoxy, that promotes spirituality through the internalisation of mysticism as the fundamental aspect of Judaism. Hasidic Judaism was established during the eighteenth century in Eastern Europe by Rabbi Baal-Shem-Tov as a reaction against the more "legalistic" method of Judaism practised by the "Mitnagdim"—the Litvish. The largest groups are the Satmar, the Belzer, the Bobover and the Chabad. Other London-based communities include the Kosov, the Biala, the Machnovka, the Chernobyl, the Skver, the Rachmastrivka, the Vizhnitz, the Karlin-Stolin and the Sassov. The various Hasidic groups emphasise the practical side of living and the communal experience, profess an emotive religion and devotion based on awe. They encouraged mystical spiritualism, intimate, active, enthusiastic, physical, vocal, spontaneous and joyful lives. The lives of the Hasidim focus around the Admor's (their religious leader) courtyard, who fulfils the spiritual role of the link between people and God, and is a central leader in the daily lives of his community. Rabbinical leadership is hereditary, usually passed on from father to son. The Hasidim speak Yiddish, and it is common to differentiate among the many different subs- and splinter-groups of the Hasidim by their place of origin. Yiddish is the primary Hasidic language, a means of identifying, socialising and differentiating the community from its surroundings. In this sense, spoken Yiddish serves to linguistically isolate the pious from the outside world.

Keywords Hasidic Judaism · Rabbi Baal-Shem-Tov · Communal experience · Emotive religion · Devotion based on awe · Mystical spiritualism · The Admor · Yiddish

Hasidic Judaism, meaning "piety", is a prominent sector within strictly—orthodoxy, comprising about half of its members, that promotes spirituality through the internalisation of mysticism as the fundamental aspect of Judaism. Hasidic Judaism was established during the eighteenth century in Eastern Europe by Rabbi Baal-Shem-Tov as a reaction against the more "legalistic" method of Judaism practised by the "Mitnagdim"—Litvish.

The various Hasidic groups emphasise the practical side of living and the communal experience and profess an emotive religion and devotion based on awe.

S. Flint Ashery, *Spatial Behavior in Haredi Jewish Communities in Great Britain*, The Urban Book Series, https://doi.org/10.1007/978-3-030-25858-0_8

Table 8.1 Rabbinical dynasties of the Hasidic groups in Stamford Hill, by subgroups

Name	Headquartered in	City/town of origin
Satmar	Kiryas Joel, New York; Williamsburg, Brooklyn	Satu Mare, Hungary (now in Romania)
Ger (er)	Jerusalem, Israel	Góra Kalwaria, Russian Empire (now in Poland)
Belz (er)	Jerusalem, Israel	Belz, Galicia, Austria-Hungary/Poland (now in Ukraine)
Bobov (er)	Borough Park, Brooklyn	Bobowa and Sanz, Galicia, Austria-Hungary (now in Poland)
Kosov (see: Vizhnitz)	Borough Park, Brooklyn; Jerusalem and Tzfas, Israel	Kosiv, Ukraine
Vizhnitz (the best known branch of the original Kosover dynasty)	Bnei Brak, Israel; Kaser, New York	Vyzhnytsia, Bukovina, Austria-Hungary (now in Ukraine)
Biala	Jerusalem, Israel; Bnei Brak, Israel; Borough Park, Brooklyn	Biała Podlaska, Poland
Machnovka	Bnei Brak, Israel	Machnovka, Ukraine
Chernobyl	Bnei Brak, Israel; Ashdod, Israel; Borough Park, Brooklyn	Chernobyl, Ukraine
Skver	New Square, New York	Skvira, Russian Empire (now in Ukraine)
Rachmastrivka	Borough Park, Brooklyn, New York and in Jerusalem, Israel	Rotmistrivka, Ukraine
Karlin-Stolin	Givat Zeev, West Bank; Jerusalem, Israel	Karlin, Belarus
Sassov	Monsey, New York; Ganei Tikva, Israel	Eastern Galicia, Ukraine
Chabad Lubavitch	Crown Heights, Brooklyn	Lyubavichi, Russia

Source Flint Ashery (2018)

They encouraged mystical spiritualism, intimate, active, enthusiastic, physical, vocal, spontaneous and joyful lives. The lives of the Hasidim focus around the Admor's (their religious leader) courtyard, who fulfils the spiritual role of the link between people and God and is a central leader in the daily lives of his community. Rabbinical leadership is hereditary, usually passed on from father to son. Although all the Hasidim speak Yiddish, it is common to differentiate among the many different subs- and splinter groups of the Hasidim by their place of origin, part of an over 200-year-old legacy (Table 8.1). For example, Viznitz, Belz, Gur and Tsanez, each one has its specific Admor, as well as the Chabad Hasidim. Hasidim usually marry others from the same Hasidic courtyard, and most of their social relations take place within it (Marcuse 1997; Peikarz 1999; Green 2001). Although individuals are free to live anywhere, households must comply with the group's directives, which dictate individuals' daily life, "from the cradle to the grave". Leavers are denounced. The Hasidim are subdivided into numerous rabbinical dynasties. The largest groups are the Satmar, the Belzer and the Bobover. Other London-based communities include the Kosov, the Biala, the Machnovka, the Chernobyl, the Skver, the Rachmastrivka, the Vizhnitz, the Karlin-Stolin and the Sassov. The Hasidic community of Stamford Hill, which will describe in the next chapter, is composed of several Hasidic groups, while the orthodox Jewish community in Canvey Island is comprised of Satmar community alone.

Yiddish is the main Hasidic language, a means of identifying, socialising and differentiating the community from its surroundings. The distribution, contents and the attitudes of the periodical newspapers in Yiddish can, therefore, shed light on the role of Yiddish among the British Hasidim. Mitchell (1999) describe that The Jewish Tribune, the only newspaper published in London itself to offer a section in Yiddish, has an overall circulation of 10,000 copies. However, Yitzhak Kraus, the editor of the paper's Yiddish section, estimates that the number of Yiddish readers is much higher as each copy rolls within the community up to its end, passes through family and relatives and to their business for the use of buyers. Der Yid, a Yiddish paper published by Satmar in New York, is the second most widely read Yiddish periodical in England. The language used is consistent with the Satmar dialect of Yiddish. Each issue consists of approximately one hundred pages, almost entirely in Yiddish, covering secular as well as religious world events. Yiddish periodicals are used to promote the ideological platform of the Haredi editors, and many Satmar papers publish articles dealing with the role of language among the Hasidic community. Mitchell (1999) argues that one reason for the lack of fundamental English language skills imparted to Satmar children is the contempt which Yeshiva young males hold for the non-Jewish content an attitude widely tolerated, if not shared by Hasidic educators. This disdainful approach to English indicates the high priority level accorded languages associated with Torah study, namely Loshn Kodesh and Yiddish. The overall appeal to reject English language books has strong emotional and nostalgic overtones. Shloyme Hertzog, the general editor of Satmar's periodical Der Yidisher Flam, explains that Satmars should preferably shun reading foreign books which can lead to assimilation, keeping to the well-beaten path of Yiddish language education. Unlike Der Yidisher Flam, which created explicitly to encourage Satmar schoolgirls to read more in Yiddish, English books with religious content do not represent the path passed down from earlier generations of rabbis and therefore constitute a "poison". In this sense, the alleged frequency of these citations in spoken Yiddish of Satmar Hasidim serves to linguistically isolate the pious from the outside world.

In the Hasidic world, based on Exodus Chapter 30 (12–13), it is common practice not to count individuals directly. For this reason, synagogue membership figures for the Hasidic community raises specific challenges as individuals may be regular attendees at a certain established Hasidic synagogue, but do not hold a formal membership, could be included in or excluded from the lists. In addition, those who own two or more memberships of different synagogues simultaneously (e.g. to be able to attend various services for weekdays and Saturdays, or to be affiliated with a "pop-up" synagogues, that operate on pay for services model) raise—in term of the research—the risk of double counting. As these dynamics within the Hasidic community pose challenging questions to synagogue data analysts, the Hasidic case studies of Stamford Hill and Canvey Island, which will be presented below, will be based on an in-depth field survey.

References

Flint Ashery S (2018) Schelling-type micro-segregation in a Hasidic enclave of Stamford-Hill. Hous Stud. https://doi.org/10.1080/02673037.2017.1414160

Green A (2001) Typologies of leadership and the Hasidic Zaddiq. In: Assaf D (ed) Zaddik and devotees: historical and social aspects of Hasidism. The Zalman Shazar Center for Jewish History, Jerusalem, pp 422–440 (in Hebrew)

Marcuse P (1997) The enclave, the citadel, and the ghetto: what has changed in the post-Fordist US city. Urban Aff Rev 33(2):228–264

Mitchell BJ (1999) London's Haredi periodicals in Yiddish: language, literature and ultra-orthodox ideology. Eur Jud 32(2):51–66

Peikarz M (1999) The Hasidic leadership authority and faith in Zadicim. Bialik Institute Publishing, Jerusalem (in Hebrew)

Chapter 9
The Hasidic Community of Stamford Hill: Non-economic Micro-segregation

Abstract The case study of Stamford Hill offers a relevant example of interactions between population groups similar in many respects, but each guarding its unique cultural identity. At the municipal level, the research area of Stamford Hill is a segregated Haredi neighbourhood, where demographic, social and economic features have heightened the growing pressure of the population on urban space in recent decades. A closer look at the inner-neighbourhood level, however, exposes the essential "micro-segregation" mechanism that takes place both at the building and neighbourhood levels. The relatively high involvement of the leadership in the community's daily life leads to a residential concentration of Hasidic communities within walking distance of the community's religious institutions and overcrowding in these areas. Stamford Hill offers lessons to the literature on segregation and the broad field of inner-city dynamics. This chapter reveals the simultaneous influence of building- and neighbourhood-level factors on the segregation pattern, which contributes to estimating the role of residential dynamics in inner cities for retaining self-identities and reproducing social segregation. It is the individuals' capacity for self-accountability that keeps the Hasidic community in Stamford Hill functioning ethically. The accepted norm whereby married women devote themselves to the socialisation of the younger generation and housekeeping places the responsibility for making a living mostly on the men. In particular, individuals are expected to organise their lives in a way that maintains religious commitments and communal identity while adapting and negotiating this adaptation to the modern reality in every aspect of their life. Within the Hasidic subcommunities, indirect collaboration within the wider Haredi society of Stamford Hill enables the Hasidic to contribute to the creation of a defined enclave that allows the conservation of lifestyles and cultures. The extended collaboration of the individuals, together with relatively weak community discipline, support and supervision, results in the heterogeneous residential pattern of the inner-city enclave of Stamford Hill.

Keywords Hasidic · Non-economic segregation · Community discipline · Supervision · Enclave · Stamford Hill

S. Flint Ashery, *Spatial Behavior in Haredi Jewish Communities in Great Britain*,
The Urban Book Series, https://doi.org/10.1007/978-3-030-25858-0_9

9.1 Stamford Hill's Non-economic Micro-segregation

The long history of geographical and sociological studies explains patterns of segregation which emphasised and highlighted the complex and continued nature of residential segregation. Studies of urban residential dynamics explain residential choice and segregation by referring solely to economic factors or by looking at a mixture of economic and non-economic factors (Borjas 1998; Clark and Davies Withers 1999; Gottdiener 1997; Mills and Hamilton 1984; Telles 1995). Non-economic factors of segregation, such as ethnic relationships, family lifestyle or life-cycle characteristics, are usually merged with the economic factors (Johnston et al. 2007), thus blurring the mechanism and impact of non-economic factors (Dudley and Roozen 2001; Dougherty and Huyser 2008). Due to the residential preferences of Haredi individuals in Stamford Hill to live among "friends", and their similar economic characteristics, the potential influence of economic factors on residential choice can be ignored, thus providing an extraordinary opportunity to concentrate on the non-economic relations maintained between neighbouring groups as stemming from their religious and cultural identity (Tausch et al. 2011; Flint Ashery 2015).

In the mid-1970s, Speare's classification identified "the socio-economic characteristics of the individual and the household on the one hand, and the socio-economic status of the housing and neighbourhood on the other, as the main factors for analysis of householders' residential behaviour" (Kasarda 1978; Speare et al. 1975). Further inquiry into these factors refined these variables; it is now common to distinguish between "revealed preferences, actual individual/household behaviour and stated preferences, individuals' declared attitudes and intentions" (Giffinger 1998; Iceland 2004; Ihlanfeldt and Scafidi 2002). As many studies show, "a comparison between stated and revealed preferences lays the foundation for understanding residential dynamics as an outcome of individuals' choices" (Benenson 2004). Particularly, the comparison of the two may reflect the roles of sociocultural motives for segregation, expressed in revealed preferences and the economic constraints, affecting the revealed behaviour (Brown and Chung 2006; Cimino 2011; Sharma 2012).

A basic approach to non-economic segregation between householders who belong to one of two ethnic groups was offered by James Sakoda and Thomas Schelling (Sakoda 1971; Schelling 1971, 1974). According to Sakoda and Schelling, "each householder considers the surrounding population to consist of 'friends', that is, householders belonging to the same group, and 'strangers'—householders belonging to other groups. Sakoda and Schelling further reduced the non-economic factors influencing the householder's decision to stay or to move to a single variable: the fraction of 'friends' within the householder's neighbourhood". According to what is now called "the Schelling model of residential segregation", "householders aim at residing in a neighbourhood where the fraction of friends, F, is above a certain threshold. In the abstract versions of a model, which consider the square grid of cells, each populating one householder only, a threshold value of F varies lays within the interval 1/4–1/3, depending on the other model parameters. This pattern explains the tendency of a householder to reside within the neighbourhood where the fraction of friends is above one-third eventually results in complete residential segregation" (Schelling 1971, 1974).

Despite the essential advance in studying Schelling model in its abstract form (Portugali et al. 1994; Benenson 2004; Flache and Hegselmann 2001; Fossett 2006), examples of the real-world dynamics that can be described by the Schelling-like rules (Benenson et al. 2002; Bruch and Mare 2006) are very few (Flint Ashery et al. 2010, 2012). In what follows, after explaining the residential overcrowding in the neighbourhood, the research considers Stamford Hill's residential pattern as driven by the interactions between householders of different groups and investigate whether the tendency to reside among people of their groups can explain residential patterns there.

9.2 Residential Overcrowding in Stamford Hill

The literature usually untangle the concepts of density according to the perspectives of "physical density—a numerical measure of the concentration of individuals or physical structures within a given geographical unit (Churchman 1999); and perceived density, namely the individual's perception and estimate of the number of people present in a given area, the space available and its organisation" (Rapoport 1975). Physical density can be view as "an objective, quantitative and neutral spatial indicator. Although there is no standard measure of physical density in town planning, it can be broadly divided into two main categories, namely people and building density. People density is expressed as the number of people or household per given area while building density is defined as the ratio of building structures to an area unit" (Longley and Mesev 2002; Flint Ashery 2018a). However, the meaning of physical density emerges from its relation with a specified scale of reference. Han and Taylor (2016) explain that "in the face of rapid urbanisation since the 1950s, densification has become an important agenda in planning policies globally. The increasing urban population and consequently the growing land scarcity pressure have attracted broad interest into the high-density development and relationship between building density and urban form, and mathematical and geometrical analyses have been conducted to address the links between building height, plot ratio, site coverage and solar obstruction" (see also Gropius 1935; Beckett 1942; Martin et al. 1972; Chizhov et al. 1968).

Perceived density, according to Cheng (2009), emphasises "the relations between individuals and the environment in addition to the spatial characteristic and the relative relationships between individual and space." Cheng further explains that adopting the two concepts of spatial density—the perception of density with respect to environmental qualities, such as high degree of enclosure, intricacy of areas and high activity levels; and social density—the perception of density with respect to the interaction between people, including the "territorial boundaries, hierarchy, the size and nature of the group involved, its homogeneity and rules for behaviour—enable us to distinguish between the different aspects of individual cognitive attributes and sociocultural norms." In this regards, a previous study that took place in Whitechapel,

London show that "While the primary problem with high spatial density, is too little space, for high social density it is too many people with whom one must interact. Nevertheless, attitudes towards high-density development are diverse, with those who acknowledge the merits of high density and advocate urban compaction, and others criticise the drawbacks and argue strongly against it" (see also Churchman 1999; Breheny 2001).

Although being a necessary antecedent of crowding, Cheng (2009) argues that "density per se is not a sufficient condition for causing the experience of crowding" (see also Stokols 1972; Baum and Paulus 1987). Churchman (1999) suggests that "unlike perceived density, which is both subjective (as it relies on individual apprehension) and neutral (as it does not involve any personal evaluation or judgement), crowding refers to the state of psychological stress that is associated with a negative appraisal of density of population in the housing unit, the consequent pressure of usage on shares facilities and lack of privacy for individual households." The case study of Stanford Hill is an example of a desirable and attractive neighbourhood where the need to rely on community institutions together with a low housing supply has led to price inflation and overcrowding. Stamford Hill has become a Haredi neighbourhood over several decades, similar to other religious neighbourhoods around the world. Stamford Hill developed gradually, and its character was influenced primarily by its central location and individual-level decision-making. The Haredi residents of Stamford Hill belong to the Hasidic groups and subgroups, which all wish to live in a strict religious environment but differ regarding their self-identity and status within Haredi society (Dahan 1999; Lupo 2004; Gonen 2006).

9.3 The Jewish Settlement in Stamford Hill

Stamford Hill (Fig. 9.1) is a district in North London consisting of four wards in the London Borough of Hackney and one ward in Haringey. Similar to other inner-city neighbourhoods around the world, Stamford Hill has developed gradually over several decades, and its character was influenced primarily by its central location and individual-level decision-making.

Interlink (2010) describes that "the Jewish settlement in Stamford Hill began as early as the eighteenth century when Jewish merchants built their country homes in what was then a rural area. In the early twentieth century, pogroms and persecution in Russia and Eastern Europe led to the immigration of 100,000 Jews into England. Gradually, Jews vacated the East End and its vicinity, moved north and concentrated in the northern neighbourhoods of the Borough of Hackney towards Stamford Hill" (see also Waterman and Kosmin 1986, 1988; Berman 2009). With the gathering of many Eastern European Hasidic Jews in the northern neighbourhoods around Stamford Hill, the cultural and social differences between "Western" and "Eastern" Jews increased. Lipman (1962: 87) explains that this process of out-migration of many middle-class Jewish residents originating in Germany "led to the closing of the schools and synagogues of various Litvish subgroups. These institutional buildings

Fig. 9.1 Location of Stamford Hill

were often taken over by incoming Hasidic groups. In 1915, the New Synagogue, which had formerly been located in central London, was moved to the area". During the 1930s and immediately post-Second World War, the Jewish population was swelled by refugees and Holocaust survivors, many of the Haredi. The newcomer immigrants were described by Blumen (2007) as a "very drawn to the advantages of the Stamford Hill area. They found an already existing Haredi population cluster, which granted them a feeling of security and provided them with appropriate institutions and services." An additional advantage was a housing surplus, given the departure of veteran Jewish residents towards northern suburbs neighbourhoods and the relative proximity to the East End area, where there were many employment opportunities.

The declaration of the State of Israel in 1948 sparked anti-Jewish manifestations in Arab lands, and the situation of Jews in those communities deteriorated markedly. While most Yemen, Iran, Iraq and North Africa Jews went to Israel in the years that followed, often in challenging circumstances, some came to Britain.

With the failed rebellion against the Soviet Union in 1956, families from Hungary joined the community in Stamford Hill, what increased the Hasidic component of the Haredi population in the area. Interlink (2010) explains that the willingness of East End's Jewry to pay higher prices for housing encouraged veteran residents "to sell and move to a newer, more spacious and fashionable house in the suburbs, and the percentage of Eastern European Haredi Jews in the Stamford Hill area increased. A few years later, and although there was no history of anti-Semitism in India, there was also an exodus of Jews from India, which had been granted independence by the British in 1947" during the 1950s and 1960s, and from the area of Aden (at the eastern approach to the Red Sea) by 1967. While the majority wished to make a new life in Israel, others came to Britain and settle in Stamford Hill. Over the years, the desire to congregate together led to the contiguous northward expansion of the Haredi population into the adjacent South Tottenham neighbourhood in the Borough of Haringey, where there had been no prior Jewish residence.

9.4 Government and Council Involvement in Stamford Hill

Although many Haredi groups in the UK "found a middle path and blended a religious life with a calculated integration into the world of modern education and employment" (Gonen 2006), most of the Hasidic men did not seek livelihoods based on career training. Those who are integrated into the workforce prefer occupations that allow them a degree of flexibility in their work schedule. Their extremely low participation in the civilian workforce together with other behavioural traditions such as young marriage and high fertility rates prevented economic improvement for the majority of Hasidic Jewish households.

Since low income and family size is the primary criteria used by the welfare state in defining eligibility, the Hasidic community of Stamford Hill is one of the most heavily supported British Haredi communities. In contrast to other cultural groups living in the same part of London, the Haredi, for the most part, do not apply for the vast reservoir of public rental housing administered by the local council, and government financial assistance provided through housing benefits provided directly to the families. Malpass (2000) describes that the "British welfare state switched from a strategy of building public rental housing to a system of direct payment of rental costs to households (see also Jacobs and Manzi 2014). The amount is conditioned upon income level and a number of children, gender distribution among children and adolescents and the permitted rental rates in each area based on market conditions" (see also Freud 2007, Department of Work and Pensions).

Rey and Folch (2011) explain that "a large portion of Haredi households in the Stamford Hill area receive economic assistance from the welfare state through child benefits, housing benefits, income support, working tax credit and assistance for certain population groups: the disabled, the widowed, and the elderly" (see also. With an average of five people per room, "child and housing benefits are the two most important programs for the Haredi population of Stamford Hill. In a survey conducted in 2001 among the Haredi community of the Stamford Hill area, it was found that 62% of households receive child benefits" (Holman and Holman 2002: 59). Holman and Holman (2002) further explain that "In the remaining households, there were no children, or the children had already grown up. The rate of households that received housing benefits was approximately 70%. The rate of income support recipients was 18%. Gonen (2006) assumes that the rental assistance received by Haredi families is in the order of 80–90% of the rental costs. This practical situation has enabled many of the Haredi residents of the Stamford Hill to remain geographically concentrated in inner-city London."

In his classic work from 2006, Gonen explains how demographic, social and economic pressures driven young families with limited incomes to establish residence far from the original group enclave. "... the generous mechanism of housing benefits, which enables the Haredi population to continue living as a concentrated community in the heart of the metropolis, is no longer sufficient to withstand the continuous rise in rental costs. The present generation of parents and in-laws is itself made up of graduates of Yeshivas and Kollels and lacks the necessary capital for financing the studies of their adult children." Despite the support of the local leadership to maintain the Stamford Hill area community and the efforts of Agudas Israel Housing Association to increase the public housing supply, the community's support is minimal, and many young families move out to the nearby suburbs or abroad. With the dissolution of the social, economic and political frameworks that constitute the background for their lifestyle, the geographic isolation that is created between the generations threatens the continuity of the community (Etkes et al. 1999; Green 2001).

9.5 Stamford Hill's Population Groups

Today, Stamford Hill is mainly known for its 30,000 Ashkenazi Hasidic subgroups, making it the largest concentration of Haredi Jews in Europe. Due to the high fertility among the Haredi communities, the area has one of highest birthdates in the UK, with what Interlink (2010) describes as "a crude birth rate of more than twenty-five per one thousand of the population—twice as high as the UK average". The household survey carried out by Holman in 2002 found that 44% of households are owner-occupiers while 54% rent their homes. Seventy-one per cent of the "Haredi renters are with private group or landlords, a significantly higher proportion than the general renting population in deprived areas, and 33% of Haredi homes were overcrowded compared to 2% in the general population". These data refer to the various Haredi residents of Stamford Hill which belong to the Hasidic groups and subgroups.

The Hasidic groups and subgroups, which all wish to live in a strict religious environment, differ concerning their self-identity and status within Haredi society. The large number of 75 synagogues in Stamford Hill reflects the diversity within the Haredi community, with different subgroups having different liturgical traditions and the need for synagogues to be within walking distance of people's homes. Previous study (Flint Ashery 2018a) differentiates among the different rabbinical dynasties and courtyards of the Hasidim: "The largest groups are the Satmar, the Gerer, the Belzer and the Bobover. Other London based communities include the Kosov, the Biala, the Machnovka, the Chernobyl, the Skver, the Rachmastrivka, the Vizhnitz, the Karlin-Stolin and the Sassov. All taking their name from the village or town in Poland, Hungary or Ukraine from where they originated, and each distinguished by some slight variation of religious practice and dress. At the head of each dynasty is the grand rabbi or Rebbe—nowadays all of whom are to be found living in Hasidic communities in New York or Israel. More than just a religious teacher, the Rebbe is held to be the fount of all wisdom and authority, on domestic, financial and marital matters—the repository of a stream of learning and wisdom that extends back through the rabbinical teachings and commentaries, to the Talmud and the Torah, and thus to God Himself. There are also negligible numbers of North African, Indian, Iraqi, Persian, Yemeni and Adeni Jews and other Haredi groups whose presence continuously declines". Each of these groups has brought its distinctive customs and traditions to enrich British Jewish life and culture. The definitions of the groups are derived from the strong self-identity of each group, also expressed in lifestyle and appearance and reflected the internal Haredi discourse.

Although the Haredi Jews are less stand out in their special physical appearance against the multicultural human landscape composition of the population in the Borough of Hackney, the case study of the Hasidic subgroups of Stamford Hill still offers an opportunity to examine non-economic processes of segregation. At the municipal level, the Haredi areas of Stamford Hill are extremely segregated within the mixed London Borough of Hackney. The research area was chosen because it is the "Hasidic Hub", an independent unit in its surrounding. Although the area is small enough and suitable for a door to door survey, its high level of Hasidic diversity reflects the Hasidic population in London. The area is populated almost solely by various Hasidic subgroups, with considerable diversity of customs and dialect. Previous studies (Alfasi et al. 2013; Flint Ashery 2018a) explain that "the residential preferences of Haredi individuals to live among "friends"—other members of the same subgroup to which they belong—together with the relatively similar economic status of the majority of Haredi families in Stamford Hill enables focusing on the role of cultural identity in the creation of residential patterns" (Flint et al. 2012). This study investigates whether the tendency of each group's members to be different from the others is expressed in their residential pattern. Stamford Hill residents share similar preferences regarding their way of life, yet powerful mechanisms of residential preferences acting at the level of the apartment and building result in "micro-segregation" patterns (Flint Ashery 2018b). The chapter examines residential relations between Haredi subgroups as reflected in their residential choices and the observed residential distribution.

The Hasidic courts are to be differentiated from the Hasidim of Chabad (Elior 1987). Chabad is a unique Hasidic movement within the Hasidic world, a dominant group at Stanford Hill. Chabad is a form of Hasidism that emphasises symbolic and ritual separation, adheres to a theology inspired by the Kabbalah (Jewish mystical writings). The latest Rebbe has not been succeeded by another Rebbe. Chabad adherent's radical claim that their latest Rebbe is the Messiah seems to put them on a different trajectory than from nearly all other groups, including Hasidic groups with similar norms and practices. Chabad is an unusually large and well-staffed organisation. Hasidic courts need to be differentiated from the Hasidim of Chabad. While many Hasidic groups are not oriented to outreach and may not be familiar with many outside attitudes, Chabad is extremely geared to outreach other Jews. Chabad's openness to modernity is a sign of its compartmentalisation (Elior 1987). Its distinct missionary vision is reflected in its demography: more than 70% of Chabad adherent were not born into the community but joined the movement later in their life. Most

Chabad Rabbis are entirely at home with the standard local language, something unusual for other Hasidim whose mother tongue will be Yiddish and their local language peppered with many Yiddish expressions. Flint Ashery (2018a) argues that "besides English, most British Chabad Rabbis speak Yiddish with a Litvish accent and grammar, unlike other Hasidim (Except for Karlin-Stolin) who speak Polish-Hungarian Yiddish. Unlike the Hasidic dress code of wearing a shtreimel and either a bekishe or a rekel, most Chabad men wear business attire with coloured shirts during the week and wear a Chabad version of the Litvish Kappote on the Sabbath". Some wear fedoras rather than the various Hasidic hats, even on Sabbath. Married women dress like Modern orthodox and wear a wig called a sheitel as they cannot show their hair in public.

9.6 The Hasidic Communities of Stamford Hill

The individuals' residential preferences of the Hasidic communities of Stamford Hill are profoundly affected by the high living costs and a competitive housing market, phenomena that lead to a spontaneous social gathering of the Hasidic communities within their parochial realm. As such, young families tend to overcrowd in the designated area of the community (see also Bastian 2011). The emergence of "nested" places—smaller communal areas inside the Hasidic area of Stamford Hill—is therefore likely to occur. The spatial-economic pressure on Stamford Hill is expressed by the activities of each community to ensure its cultural and social institutions. Similar to other communities in the Stamford Hill area, the Bobov Hasidic group is "making an effort to secure its economic base in Inner London. As part of this effort, they preserved the property of the Jewish population by not joining the great migration of Jews to the suburbs and made use of the social investments in Jewish institutions in the area made before their time" (Interlink 2010). Thus, they have purchased the New Synagogue building and school buildings located near the synagogue from the United Synagogue (central orthodox) stream and established a compound of its communal institutions.

Gonen (2006) describes the unique "expenses incurred in the Haredi lifestyle and mainly the consumption of kosher food, the frequent long-distance travel in order to participate in family events, costs for sons who learn in distant Yeshivas and expenses for a respectable dowry for daughters, even if the prospective son-in-law does not plan to be a full-time Kollel scholar". The highest cost is of Haredi education which is not covered by the state: "The cost of these institutions in the Stamford Hill area is particularly high due to the multiple divisions between Haredi groups, each of which wants to run its educational institutions. There are at least fourteen Talmud Torah schools in the area, which constitute the basis for elementary school education for boys in the spirit of the Haredi culture. Most of the Talmud Torah schools are Hasidic, due to the present make-up of the local population: Bobov, Belz, Vizhnitz, Satmar, Lubavitch and Skvera. But there are some additional such Haredi schools in the Stamford Hill area not related to any specific Hasidic group, the most prominent of which is the Yesodey Hatorah Talmud Torah, that caters to a wide spectrum of Haredi groups. Due to their refusal to adhere to the National Curriculum stipulated by the British Ministry of Education, these educational institutions are not eligible for state support. As a result, the two layers of Yeshivas beyond the age of Talmud Torah, the "small" and "big" layers, are run by some of the local Haredi groups in the Stamford Hill area, and all contribute to the increased financial burden on Haredi households. Added to these is the separate track of girls' schools, also funded by parents. Only of late has there been some relief, following the State's recognition of Yesodey Hatorah School for Girls operating in the area for several decades".

"Interlink" (2010) also explains that Haredi is "continually engaged in adapting and extending the housing stock to meet their needs. This is true of poorer Haredi households as it is of the more affluent members of this classless community. Despite the high level of enforcement, there are hundreds of loft extensions, dormers, back extensions and other adaptations across the neighbourhood". However, the prices of apartments in Stamford Hill, both rental units and units for sale, are relatively high. This is an expression of the area's attractiveness and reflects its desirable inner-city location, the wide range of religious institutions and rabbinical courts active in the area and the population's high status. In what follows, the chapter introduces the different groups living in Stamford Hill, characterises their residential interactions and analyses their residential behaviour.

9.7 Methodology: The Case Study of Stamford Hill

To investigate the bottom-up residential relationships among the research area of Stamford Hill's population groups, the study uses a mixed methods approach to construct a detailed spatial-temporal database that contains accurate geo-referenced data on Stamford Hill families' religious affiliation and to analyse data from this extensive bottom-up micro-resolution. The field research was conducted during 2015 at the level of individual families and apartments. Two interviewers, young Haredi males, canvassed the neighbourhood and conducted a door-to-door survey. They asked all of the Stamford Hill research area's occupants to identify their Haredi group as well as the apartment's former dwellers, going back to 2000. Several researchers stress "that the identity of the previous residents is important for Haredi families" (Waterman and Kosmin 1986), a conclusion confirmed by this research. Identification of past residents allowed us to understand which group's members had occupied each apartment for the past 20 years. All other questions asked related to the present occupants to ascertain their socio-spatial behaviour.

The interviewers collected data about the location of the synagogues the families attended, whether the apartment was owned or rented, and the source of information about the apartment before buying or renting it. In cases when the householder refused to cooperate, the interviewer obtained information about the occupant's group by questioning neighbours. Despite early apprehensions regarding cooperation, the response rate reached 83%. Upon collecting data, householders were also questioned about their motives for choosing the apartment and asked to rank the relative importance of the apartment's price, their neighbours' identity and institutional proximity.

Construction of the Stamford Hill GIS was based on the layers updated for the year 2015, which were provided by the ordnancesurvey.co.uk/opendatadownload/products.html. The characteristics of all the research area's apartments and households were organised as a GIS layer, in which every record in the table is related to the corresponding building. Since most of the buildings are private houses, mainly typical British semi-detached, the collected data are conventionally present for every 8–10 apartments, to protect the confidentiality and keep a building-scale resolution. The layer was then included in the area's high-resolution GIS. Stamford Hill GIS contains additional layers of topography, roads, land parcels and buildings, and the latter characterised by use and number of floors.

In-depth Interviews: As coresidency with the extended family and dividing a house to a few flats are so common, data collection raised practical challenges of dividing the flats and analyse the data of each one of them without publishing the results on maps. Participants were accessed directly and ensured the variety and diversity of households. Interviews were carried out between February and May 2015 by experienced fieldworkers with relevant language fluency for each community and myself. I have chosen 24 women and 19 men from different parts of the neighbourhood that represented various age (20s–70s), family situation, different Hasidic affiliation and social location that emerging as factors in their residential choices. The interviewees were asked about their residential opportunities, daily life, inter- and intra-group relations.

The research area of Stamford Hill contains 104 apartments in seven buildings with another 686 apartments in 406 detached/semi-detached houses. A total of 1165 households occupy 590 flats. Institutions use another 67 apartments (such as synagogues), and the remaining 71 are shared by households and institutions (such as kindergartens operated within dwellings). Taken as a whole, the survey's spatial-temporal GIS enables evaluation of residential patterns at the resolution of apartments, buildings and neighbourhood; it makes an investigation of the residential micro-dynamics in this limited environment empirically possible.

9.8 Estimation of Residential Segregation

Following several other studies in this field (Anselin 1995; Zhang and Lin 2007), "the level of spatial segregation at the resolution of buildings was estimated with the Moran I index of spatial autocorrelation. The Moran I index was applied for evaluating the correlation between the fraction of a given group D in building I and the fraction of D over the buildings $U(i)$ that are adjacent to I", as we can see in figure.

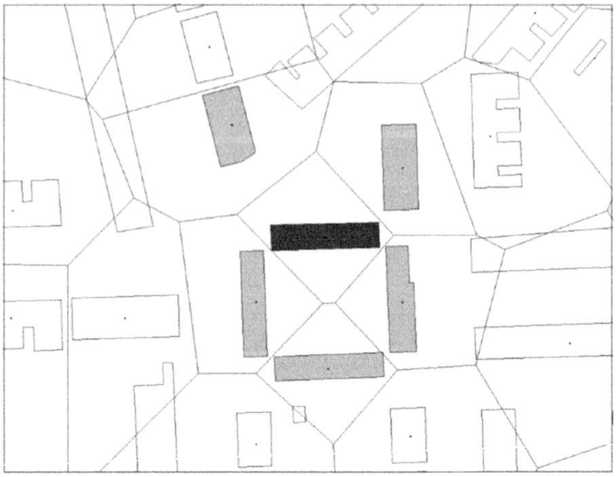

Fig. 9.2 Stamford Hill buildings and the coverage of Voronoi polygons constructed based of buildings' centroids. Voronoi-based neighbours of the selected building (in black) are shown in grey. *Source* Flint Ashery (2018a)

$$\frac{N\sum_i \sum_{j\in U(i)} w_{ij}\left(D_i-\overline{D}\right)\left(D_j-\overline{D}\right)}{\left(\sum_i \sum_{j\in U(i)} w_{ij}\right)\sum_i\left(D_i-\overline{D}\right)^2}$$

An estimation of the correlation between the fraction of a given group D in building i and the fraction of D over the buildings $U(i)$ that are adjacent to I as applied by the Moran I index.

Where N is the number of buildings and $\overline{D}$ is the average fraction of group D in Stamford Hill. The influence w_{ij} of the neighbouring buildings $j \in U(i)$ on i is calculated as $w_{ij} = 1/NU(i)$, where $NU(i)$ is the number of buildings in $U(i)$. The proximity of buildings defines by a Voronoi partition constructed by the buildings' central points, as proposed by Benenson et al. (2002). According to this definition, two buildings are adjacent if the central points of their foundations are directly visible by the other (see Fig. 9.2):

According to the collected data, the population in the research area of Stamford Hill is growing, initially populated by various Hasidic group, with a non-Hasidic group (mainly North African, Persian and Adeni Jews) steadily replaced by Chabad residents during the late 2000s (see Fig. 9.3a–c). Arele (pseudonym, male, 37) describes the changes in the neighbourhood: "Stamford Hill has been identified for many years with the Haredi population. But the Haredi community is also changing. The number of Hasidic families living in the neighbourhood has increased, and the number of children in each family has increased, and we all need to live together. Today, if you have less than six children, people are raising an eyebrow over it. In the past, there were families of non-Ashkenazi Jews who had fewer children. Now they left. The problem is that as the neighbourhood grows, especially in the direction of South Tottenham, there is more friction with the population there".

Currently, the area is regarded as exclusively Hasidic. Despite the tight organisation of the Haredi sector, Stamford Hill's housing market is formally free, with people buying or renting apartments in accordance with their personal preferences

Fig. 9.3 a–c: Dynamics of spatial distribution of the non-Haredi versus Chabad. The patterns are for 2005, 2010 and 2015

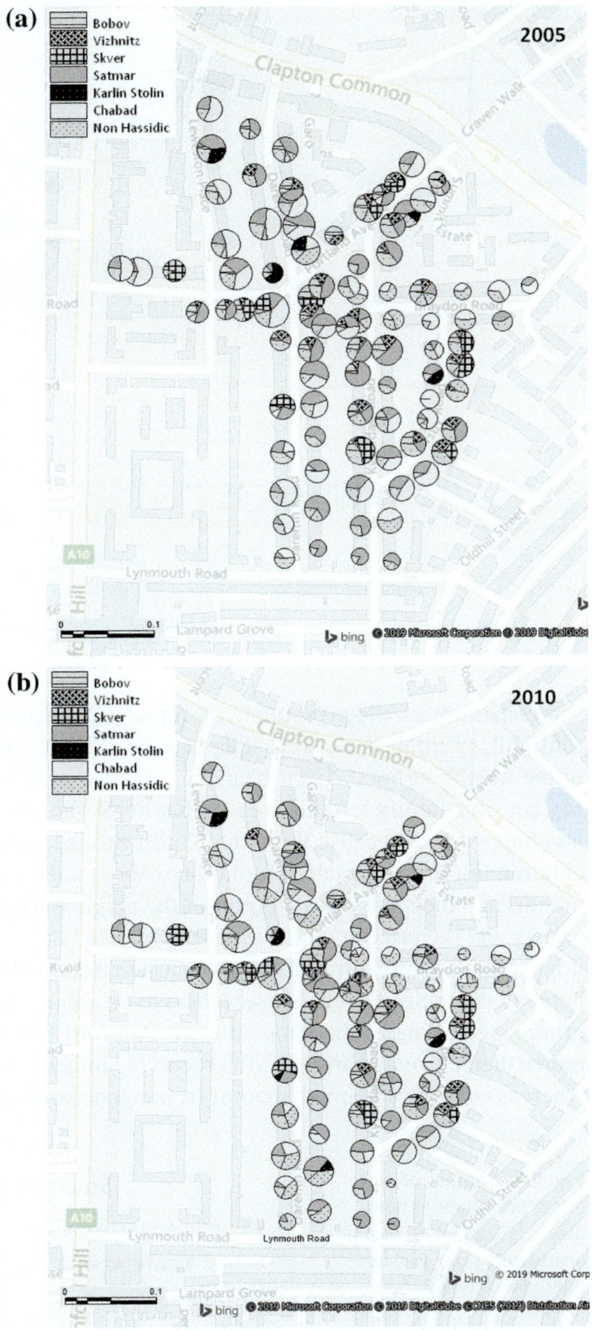

Fig. 9.3 (continued)

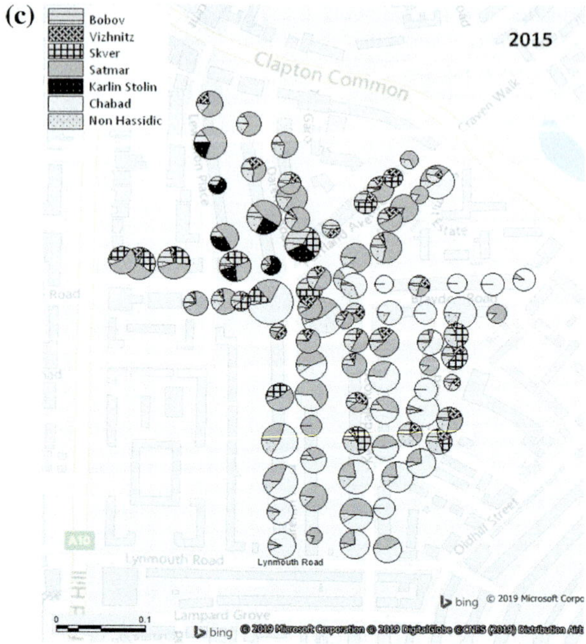

and constraints. Golde (pseudonym, female, 25) describes the free market in Stamford Hill: "I think if we were able to afford it, we would be able to buy anywhere in the neighbourhood. Even near the Yeshiva, even where the Karlin Stolin lives. My cousin's family left this area (of Karlin-Stolin), and his wife said that they were less happy there because they (the Karlin-Stolin) were very much together and her children felt uncomfortable. I told them about a flat that is available here and a few months ago they moved to this part of the neighbourhood because here there is more variety and the price at which they sold allowed them to buy a bigger flat with additional room". Close intra-group relations are responsible for the efficient information network that helps Haredi householders learn about vacancies from members of their group. According to the data, 73% of Stamford Hill's population either bought or rented their current apartment after receiving information through family and friends, whereas only 25% learned about it from agents and advertising (see Fig. 9.4). Note

Fig. 9.4 Sources of real estate information: how dwellers learned of vacant apartments prior to their purchase or rental

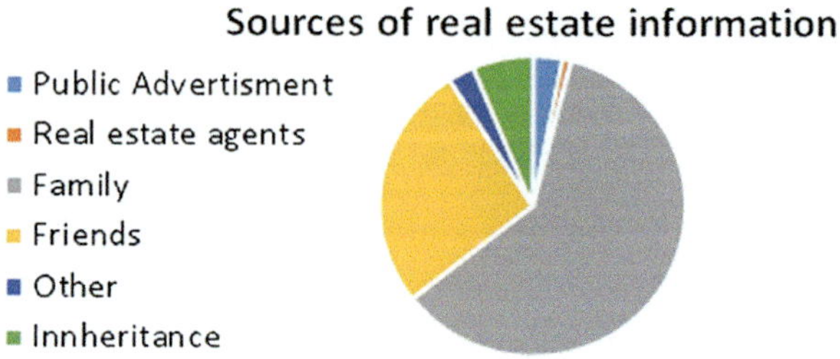

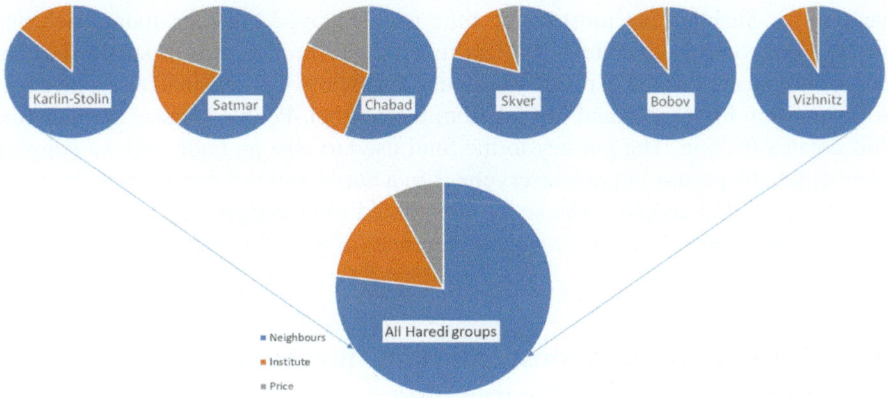

Fig. 9.5 Importance of apartment cost, neighbours' identity and proximity to institutions in apartment choice by Haredi group, Stamford Hill (2015)

that the distribution of information sources does not depend on affiliation to groups (chi-square test, $p \sim 0.4$).

Figure 9.5 shows that Haredi groups in Stamford Hill share similar concerns (chi-square test, $p \sim 0.5$). Less than 26% of Stamford Hill dwellers chose the location of Haredi institutions as their primary concern. Stamford Hill's location close to the city centre ensures the proximity of such institutions. Contrary to economic theory, only up to 18% of each group indicated that price was a critical issue for them. Most important, however, is the fact that, despite the neighbourhood's reputation as a Haredi neighbourhood, more than 77% of Stamford Hill dwellers reported that the identity of their immediate neighbours is their principal concern. As this stated preference appears to be shared by members of all groups, the research can assume that the Schelling-like mechanism of actively distinguishing between "friends" and "others" remains relevant within the Haredi haven of Stamford Hill. Most of the neighbourhoods' dwellers feel the need for at least a few "friends" to feel at home in their apartment building.

Aver (pseudonym, male, 70) lives all his life in Stamford Hill: "Living with the community is good in every way: you always have support, if a person is suddenly ill or he is alone, he can go to the synagogue or go to the neighbours. If you walk down the street, there will always be someone to ask for your welfare. A few weeks ago I was unwell and couldn't come to the grocery store for two days, and the baker brought here two nice Chullas for Shabbath without asking for payment, but only because my neighbour told him that I was not feeling well … Every time you come to the Shul (synagogue) there will be a Minyan, you are always surrounded and protected. It's like that now, but even when we raised children, it was like that. There is Gemach (an interest-free Jewish fund operates in most Jewish communities for lending goods on a short- or long-term basis) for everything that is needed, and people will only ask to do charity and help … Back then, there was a synagogue for every Nosah (a specific version of praying), and every small community searched for its distinctive version for Shabbat, but most of the time we all prayed together, today each tiny group want to pray separately. The entire neighbourhood is full of small Shtibels, some of them are located in residential council flats, and it is now that everyone wants to live within his people and not far from the Shtibel. People became spoiled, and this creates friction. The journey to the Shul used to take an hour, and we enjoy it, also with baby prams, but now everyone is in a hurry, and this creates wars between groups of spoiled people". The study thus turned to investigate the impact of these declared preferences on the revealed preferences of Stamford Hill's dwellers.

9.9 Bottom-up Initiations: Interflat, Interbuilding and Intra-building Segregation

The tendency for coresidency in Stamford Hill's apartments varies between different ethnic groups. Table 9.1 shows the mean and median of the number of families living in apartments that are occupied by different Hasidic groups. Dovid (pseudonym, male, 31) of Kosov describes: "Density like the one we keep has a lot of advantages. One of them is the close supervision of everyone on everyone all the time. Little children share a room with all their sibling, boys with boys, girls with girls and that's how they grow up when they learn to share". The study reveals that Satmar families have the highest tendency to share apartments. Rabbi Pinter, a Satmar rabbi and politician who lives in Stamford Hill, explains: "The main reason for Satmar to share flats is that we all want to live in this area, each family has several children, and when the prices are high, the parents find it challenging to meet the needs and invite the families of their children to live with them. They close a room and turn it into a flat, so there is no problem of modesty. And then another room for the next child and so on. The problem is that most of the families do it. Even the wealthiest. *G*d* enabled me to allow all my children to live in the neighbourhood, and now I

Table 9.1 Number of families in apartments by Hasidic groups

Group	Mean	95% CI	Median
Satmar	2.99	[2.85, 3.14]	3
Belz	2.37	[2.04, 2.70]	2
Bobov	2.28	[2.06, 2.50]	2
Ger	1.72	[1.54, 1.89]	1
Vizhnitz	1.60	[1.25, 1.95]	1
Skvera	1.50	[1.38, 1.62]	1
Karlin-Stolin	1.49	[1.25, 1.73]	1
Chabad	1.47	[1.28, 1.66]	1

The groups are listed in descending order according to the average. *N* is the number of apartments in each group which occupies in the survey am taking care of the next generation of our grandchildren. So yes, we may live in a higher densely level than people live in Finchley or Golders Green, but that's good for us. We want the space to be like that. We make it be like that. If you take a Satmar neighbourhood in Israel, in Bnei Brak or Mea Shearim, it will not be like that? It will" (20 December 2017).

Also, the survey data enabled reconstruction of Stamford Hill's residential patterns for the entire period between 2000 and 2015 (see Fig. 9.6a–c). Despite the stated preference for living in a "friendly" environment, those maps indicate a high level of spatial integration, with members of several subgroups sharing apartment buildings and near neighbourhoods.

Moran's I index appears relatively high for two groups (see Fig. 9.7). The dynamics of Moran's I index show that the Chabad and Satmar Hasidim are the most highly segregated group in Stamford Hill, although the residential segregation of the other groups has been steadily growing over the years. Shaina (pseudonym, female, 33) of Satmar explains the Satmar's tendency to be relatively segregated: "All of our men work within the community, work and study and do Chesed (the attribute of benevolence or compassion). Satmar has always been doing Chesed and give charity within the community and outside the community. We have large organizations that help patients in hospitals and their families all with professional medical assistance. Anyone who needs help knows how to get here, but we prefer to be isolated. We live in Stanford Hill for generations now and all our lives concentrate here. So when everyone here, together with our Torah lifestyle, studies, Chesed, work, children—it encompasses all aspects of life within this small and crowded area where everyone lives together and we can raise our children. So from outsiders' point of view five people per room may sounds a lot, that there is no privacy. But it's part of life when we all live together, with no distractions from others. And there is supervision of God and the community".

A high-resolution looks into the dynamics reveals the relationship between the stated (see Fig. 9.5) and the revealed residential preferences of the area's residents. To estimate the latter, for each group D the study calculated the fraction of D-families in each building that had a vacant apartment. Then, for each group D, the study now compared the average fraction of D-families in buildings where the empty apartment of which was chosen by new D-families, to the fraction of D-families in the rest of the buildings. For each group D, the average fractions of friendly families in the buildings chosen and not chosen by D-families are presented in Table 9.2. The buildings selected by D-families show percentages of "friends" which are up to two times higher than the buildings with vacant apartments ignored by D-families. This fundamental pattern of residential behaviour is consistent for all religious groups in Stamford Hill, although weakening at the near neighbourhood level (see Table 9.3).

Note that although the relatively small number of Karlin-Stolin members is insufficient to express their segregation tendency by the Moran's I index (see Fig. 9.6), a comparison between the average fraction of families in buildings the vacant apartment of which was chosen by new families and the fraction of D-families in the rest of the buildings indicates that they effectively implement their stated preferences by residing in buildings occupied by "friends". Beynish (pseudonym, female, 18) of

(a)

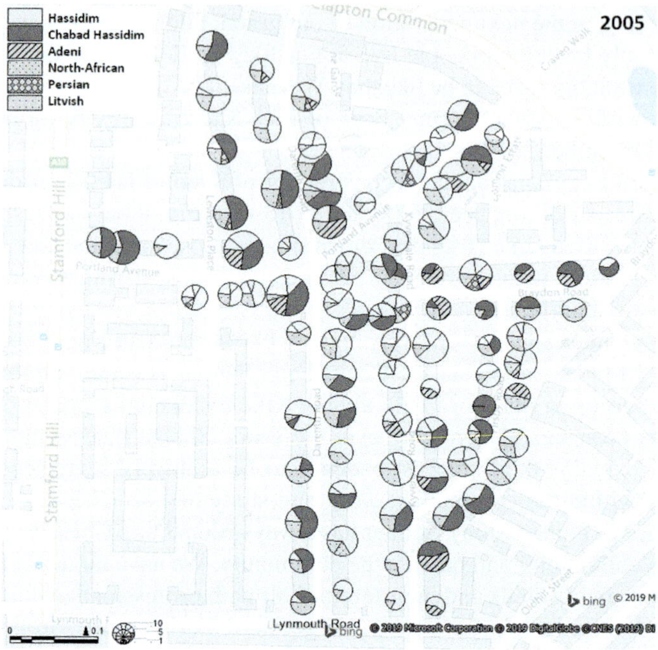

(b)

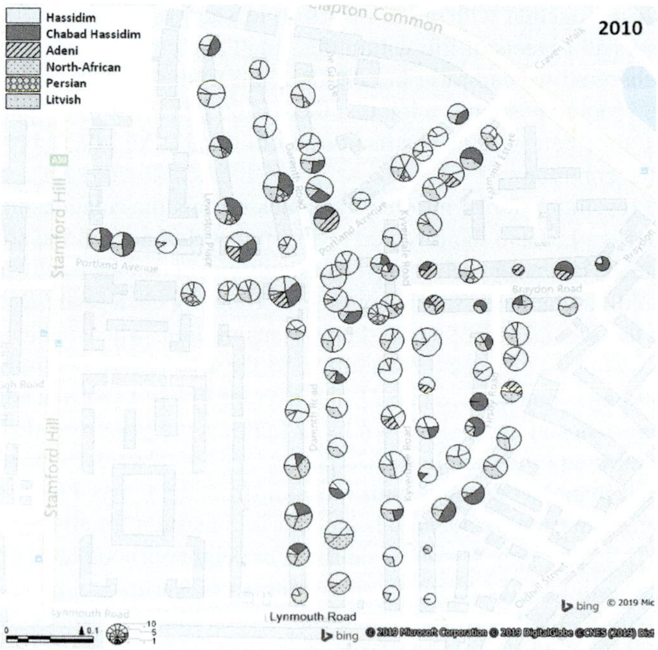

Fig. 9.6 a–c: Spatial distribution of Bobov, Vizhnitz, Skver, Satmar, Karlin-Stolin, Chabad and non-Haredi Jewish groups in apartment buildings, Stamford Hill 2000–2015

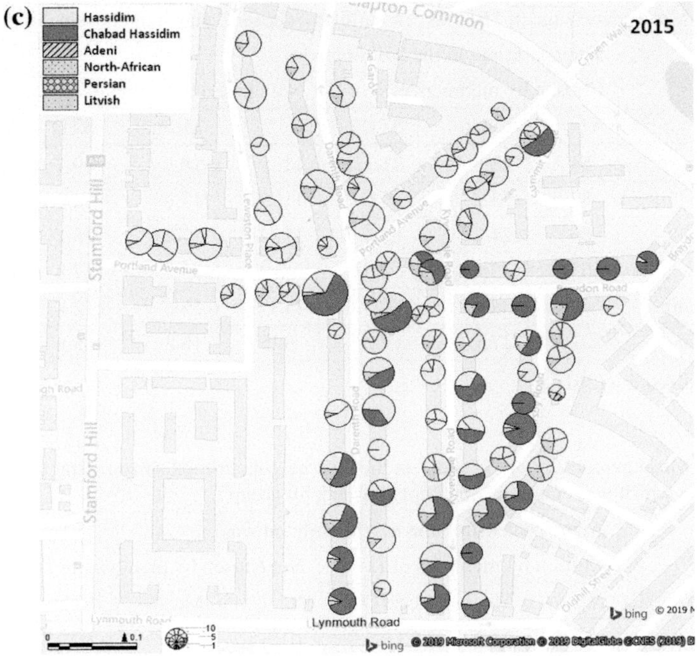

Fig. 9.6 (continued)

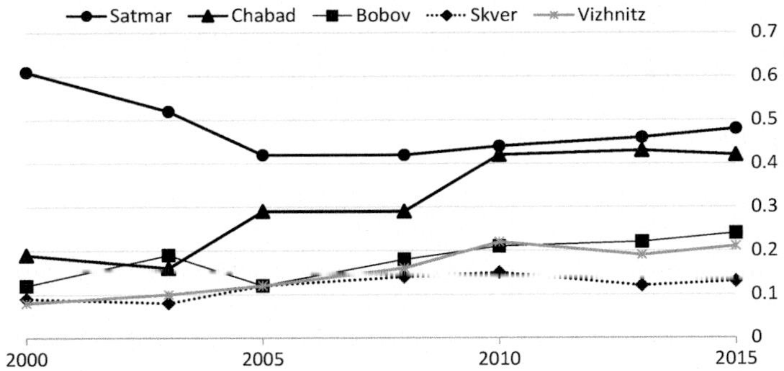

Fig. 9.7 Segregation of Stamford Hill's Hasidic groups as expressed by the Moran's *I* index of spatial autocorrelation, during the period of 2000–2015. To be significant at the 5 or 1% levels, the value of the Moran's *I* should be above 0.1 and 0.15, respectively

Table 9.2 Mean percentage of "friends" in building "chosen," and "not chosen" by group

Group *D*	The apartment in the building was …						*P*
	Chosen by members of *D*			Not chosen by members of *D*			
	N	Mean	STD	*N*	Mean	STD	
Bobov	54	28.6	21.3	1128	18.4	19.7	<0.001
Vizhnitz	57	25.2	16.6	1125	17.6	16.2	<0.001
Skver	60	24.2	17.4	1122	17.7	16.1	<0.001
Satmar	663	35.8	32.6	519	13.2	17.8	<0.001
Karlin-Stolin	47	38.4	11.4	1135	15.6	12.4	<0.001
Chabad	207	27.9	17.2	975	16.8	15.8	<0.001

Source Flint Ashery (2018a)

Table 9.3 Mean percentage of "friends" in the near neighbourhood surrounding the buildings the apartment in which was "chosen" and "not chosen", by group

Group *D*	Apartment within the neighbourhood was …						*P*
	Chosen by members of *D*			Not chosen by members of *D*			
	N	Mean	STD	*N*	Mean	STD	
Bobov	52	24.7	15.5	1018	9.8	10.2	<0.001
Vizhnitz	56	22.1	9.7	1014	7.6	9.1	<0.001
Skver	58	20.1	10.2	1012	8.2	10.4	<0.001
Satmar	659	37.1	22.1	411	14.7	12.3	<0.001
Karlin-Stolin	47	32.9	14.7	1023	21.2	25.4	<0.001
Chabad	198	24.6	13.5	872	15.2	11.8	<0.001

Source Flint Ashery (2018a)

Karlin-Stolin briefly explains that "It is highly important for us to stay away from the others because they do not accept our custom to cry out loudly to God when praying. That is why we live together". Anshel (pseudonym, male, 31) of Bobov provides another explanation for this tendency: "Everyone is tense here about properties. This is our future, of our children and of the community. When you hear about a flat that is vacated, everyone hears about it from their family, or other Anshi Shlomeno (People who have a similar conceptual identity, a Hasidic nickname for people in the same social group who can be trusted) in the synagogue after the service, many times they close deals in the Shul. If there is a Bobov (flat of someone from Bobov) I'll hear about it in Bobov's circles, so Bobov will buy it. There are also links between the Hasidic groups thus information goes out as well. But usually, there is a delay so when there is a new flat comes to the market the deal closes quickly within the community. I know that that's the way elsewhere. When you buy a flat from inside, you get not only the flat but also the neighbours, the places at school, the proximity to the Shul. You know that the kitchen will be organised exactly as it should be". As these preferences appear both in the stated and in the revealed preferences, it is safe

Table 9.4 Fraction of members of a given group who occupied an apartment vacated by a member of their own group

Group	Bobov	Vizhnitz	Skver	Satmar	Chabad
Fraction	0.92	0.94	0.81	0.85	0.52
N	213	234	152	541	552

Source Flint Ashery (2018a)

to conclude that they form a fundamental mechanism, providing people with a sufficiently strong sense of home and belonging (Flint Ashery 2018c). This study traces the process of organising Stamford Hill's residential patterns throughout the years and reveals the central role of the parochial realm inside the Hasidic area (Lofland 2009). A high percentage of Haredi families reside in apartments vacated by householders of their group (see Table 9.4). The high probability of an empty apartment being transferred to members of the same group becomes a parallel, individual-based mechanism supporting micro-congregation. Families can thus be assured that the level of "friendliness" in their building will not decrease.

9.10 Intergroup Relations

The Schelling-like tendency to reside in buildings and neighbourhoods with a sufficiently high fraction of friends explains the evolution of residential patterns in populations consisting of two groups. This model, however, is insufficient to explain the case of Stamford Hill, where householders from several groups compete for the same residential space. Moreover, in contradiction to Schelling's symmetric view of residential relationships, relations between Haredi groups are hierarchical. It is well-known, for instance, that Hasidim consider proximity to Chabad to be embarrassing (Berman 2009; Friedman 1991). One can assume that Hasidic householders would avoid residing in a building with one or more Chabad families, although Chabad householders may be tolerant towards Hasidic neighbours and even prefer them. It is also known that despite the differences between the Hasidic courts, there are marriage relations between other Hasidic groups. Which, if any, of the relationships between the groups, are relevant for explaining residential patterns in Stamford Hill?

To estimate the residential attitudes of members of group *D* towards members of other groups, the research compared the residential structure in buildings with vacancies that were/were not selected by householders belonging to *D*. If two buildings have the same fraction of families belonging to group *D* but different fractions of families of other groups, and *D*-newcomers prefer/avoid residence in one of those buildings, a comparison of the fractions of the other groups residing in these two buildings should reveal *D*'s preferences in regards to the other groups in the building. This procedure was applied when the steady out-migration of non-Haredi householders from Stamford Hill was almost completed (see Fig. 9.3a–c). The analysis begins with comparing "buildings of others" in which *D*-families did not reside at all at the moment of residential choice.

Satmar families are indifferent to the identity of others in the building. This is not so for Bobov, Skver, Vizhnitz and Karlin-Stolin householders. Satmar, if forced to choose apartments in "buildings of others," surprisingly chooses houses with high fractions of Adeni and Chabad families. Here, for the first time, the indirect impact of social status plays: as the Satmar presence spread in Stamford Hill during this period, their social status provided them with the power to select the most desirable locations. Faivish (pseudonym, male, 59) of Satmar describes: "I think everyone here appreciates Satmar's contribution, both to the Torah world and the image of the Haredi in the eyes of the general population. This is, of course, reflected in the fact that if a flat is vacated and there are families from other groups that give offers in a similar price, there is a good chance that the offer which was presented by our family (of Satmar) will be accepted because everyone knows what does it means to deal with Satmars. Our families are also confident enough to live solely in the building of another group, even with the general British population, even with non-Jews". Satmars, appear to prefer buildings with less competition, although the non-Haredi residents in those buildings belong to groups "below them" in the neighbourhood social hierarchy. When "friends" reside in the building, all besides Bobov, Skver and Vizhnitz newcomers become indifferent to the proportions of others. They remain sensitive to the building's population structure and avoid/are deterred from buildings with more Chabad, but are attracted by more Satmars. Chabad families, however, prevent or buildings with high fractions of Bobov, Skver, Vizhnitz and Karlin-Stolin and prefer or are steered to buildings with high fractions of Satmar and Adeni families.

As previously shown (Flint Ashery 2018a), this is another manifestation of the critical role played by the social hierarchy in the formation of residential structure: "individuals belonging to higher-status groups are in effect less concerned by the identity of others in the building, as long as there is at least one family of 'friends'. However, households with lower social status are much more limited in residential choices and end up with a greater presence of 'friends' in an apartment house. In conclusion, individuals and households are sensitive to the identity of others in the buildings, and when they cannot find 'friends' as close neighbours, they choose their residents according to the social hierarchy. Individuals from marginalised groups associate with households of other marginalised groups, whereas individuals located at the top of the social hierarchy are in effect free to choose their neighbours from all other groups. In the case of Satmars, they can express their high social status by choosing cheaper apartments, located in buildings with non-Haredi households".

The last issue concerns the near neighbourhood (excluding the building itself). Comparing chosen and not-chosen buildings, the study compared group percentages within the adjacent neighbourhoods of buildings that had a vacant apartment. Only Chabad was found to be sensitive to the group distribution of the near neighbourhood in addition to the building and tend to avoid/are steered away from other Hasidim but the Satmars. In total, the need to live close to someone who shares the same faith, worldview and lifestyle is shared by all the groups examined and the urge to avoid those who are "not friendly" is partial and weak.

Another interesting finding concerns the asymmetric nature of segregation attitudes. Chabad and the four Hasidic groups of Bobov, Skver, Karlin-Stolin and Vizhnitz—relations maintain some one-way relations: "Chabad avoid buildings where the percentage of those Hasidim is high, yet the Hasidim are less sensitive to the portion of Chabad; Satmar are insensitive to the fraction of Chabad families in the building while Chabad prefer buildings with a higher percentage of Adeni. Resulting from the lack of symmetry, different groups face different limitations when choosing a new building. While individuals from higher up in the social hierarchy focus on the building and tend to neglect the near neighbourhood, households of lower social status are more sensitive to the distribution of their group in the near neighbourhood, as a compensation to the relative absence of a 'friend' in the building, and are usually forced to choose apartments with a higher fraction of non-Haredi households".

The current behaviour described above is coupled with other tendencies that have shaped the spatial configuration of Stamford Hill throughout the years. For decades, Chabad families have concentrated around their institutions, on the northern area of Portland Avenue and along Stamford Hill Road, a major highway considered less attractive than the neighbourhood's interior areas. Recently, Chabad has concentrated on the southern area of Portland Avenue (see Fig. 9.3a–c). All this leads to the assumption that the residential preferences of Chabad families manifest feelings of social exclusion rather than a free choice between market alternatives.

Exclusion of Chabad from buildings is especially striking when compared with the segregated pattern of Karlin-Stolin. From the mid-1990s, Karlin-Stolin began to make themselves felt in the stronghold of the "Hasidic Hub". While they rarely managed to purchase apartments in the area, by 2015, 73% of Karlin-Stolin continues to rent from the Chabad owners (see Table 9.1, bottom row). Regarding the apartments vacated by non-Haredi families, the Satmars' competitors are the Chabad. Chabad is taking advantage of their economic abilities and are the first owners and renters of the apartments vacated by North African, Persian and Adeni families. The fraction of non-Haredi families in the "buildings of others" chosen by both groups is essentially higher than the respective percentages in buildings not chosen. However, we can assume that the occupation of these buildings by Satmars expresses their superior status and freedom of choice, whereas Chabad resides there as an expression of their social exclusion and lack of choice. In conclusion, this intra-group and intergroup analysis has revealed mechanisms of spatial differentiation through the congregation with "friends" while selectively ignoring unwelcome "others". Since these mechanisms operate more strongly on the building than on the near neighbourhood level, it is appropriate to refer to the aggregate results under the heading "micro-segregation" (Flint Ashery 2018b).

Moreover, the tendency to congregate within the same area which also maintains a relatively high level of crowding and increases the number of families per building. Figure 9.8a, b compares the actual number of families in the buildings with the number of flats in the building. Although there are groups that are able to maintain higher density than others, the overall result in of overcrowding (Fig. 9.8a, b). A comparison between 2005 and 2015 reveal that the level of density per building increases. The urgent need for housing solutions for the population that wants to live in the

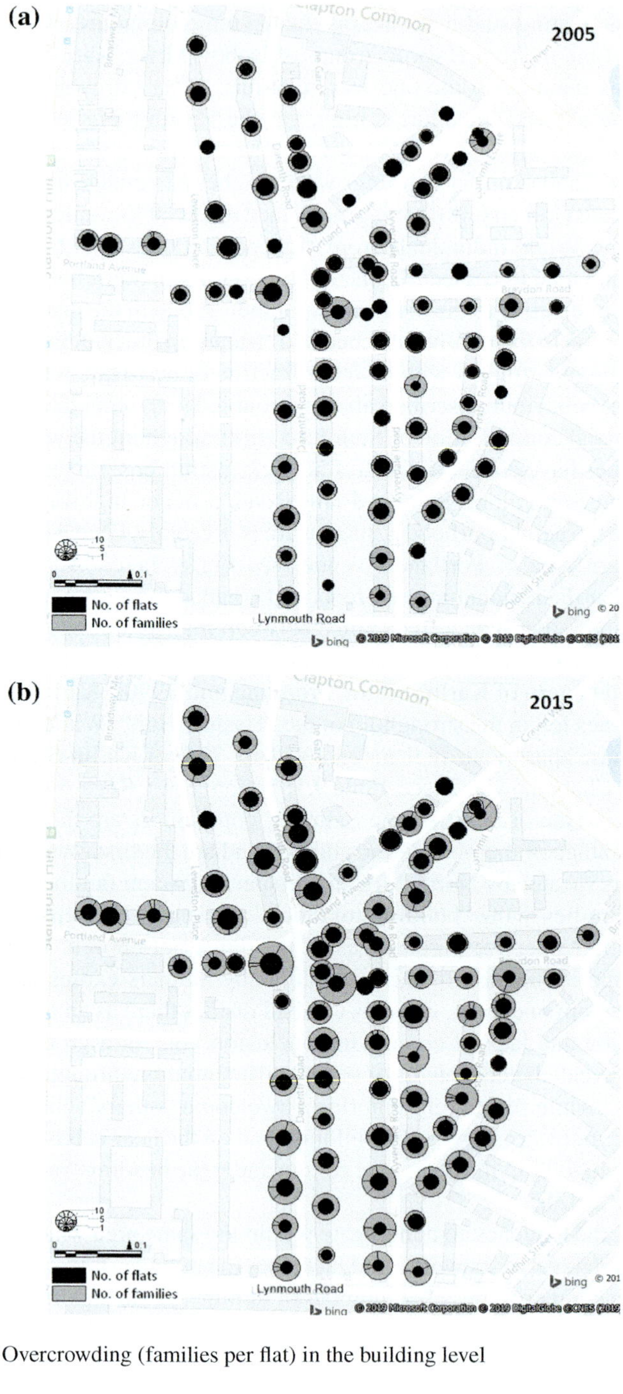

Fig. 9.8 **a**, **b**: Overcrowding (families per flat) in the building level

same area, to live between similar neighbourhoods and rely on community institutions leads to many building irregularities, mainly loft extensions, dormers, back extensions and other adaptations across Stamford Hill (Fig. 9.9a–e). Such strategies among dwellers are reinforcing the physical parochial realm and contribute to urban compactness.

Fig. 9.9 Loft extensions, dormers, back extensions and other adaptations across Stamford Hill

9.11 Discussion: Local Residential Preferences as a Driving Force to Non-economic Segregation in Stamford Hill

This study examines the impact of social relations and social identity on intra-urban segregation, focusing on the research area of Stamford Hill, an inner-city neighbourhood in London. The literature assumes that non-economic forces, namely communal relations and social considerations, are the main driving forces of internal neighbourhood segregation (Peach 2006). This is especially true for Stamford Hill's Haredi population. Generally, relatively intense devotion to lifestyle leads the Haredi population to focus on group affiliation and enforce residential homogeneity despite turnover in the buildings and neighbourhoods. Once economic considerations are set aside, Stamford Hill householders are free to direct their attention to inter- and intra-group relations and emerging social patterns when choosing a residence. At the municipal level, the research area of Stamford Hill is an extremely segregated neighbourhood, occupied almost solely by Haredi inhabitants. A closer look at the inner-neighbourhood level, however, exposes essential "micro-segregation" that takes place at the building and neighbourhood levels. The Stamford Hill case study offers a relevant example of interactions between population groups similar in many respects, but each guarding its unique cultural identity. The unique information collected via in-depth door-to-door survey made it possible to compare stated and revealed preferences at the household level and to reveal powerful migration mechanisms that govern this segregation over two decades.

The religious affiliation of the building's inhabitants proves to be the motivating force for dwellers from all groups. Non-Orthodox people living in Stamford Hill are leaving because Haredi from the two poles of the local social spectrum—the high-ranked Satmar and low-ranked Chabad—quickly occupies their vacated apartments. The Hasidic group of Karlin-Stolin plays the role of the most highly segregated group, most anxious to congregate with one another. Other Hasidic groups of Bobov, Skver and Vizhnitz avoid specific others in the building level, such as the Chabad. Chabad is the only group paying almost the same amount of attention to the number of "friends" in a nearby building as in their own. Chabad is socially excluded from other Hasidic-dominated areas. Whether Chabad tend to avoid other Hasidim or feel unwelcomed by them, they are nevertheless more likely to move in with the marginal non-Haredi and Satmar.

Stamford Hill is a laboratory for studying micro-segregation and offers lessons to the literature on segregation and the broad field of inner-city dynamics and invasion–succession processes. To begin with, the research findings extend Schelling's theory regarding the symmetric relations between two groups by adding two essential points. First, instead of Schelling's "white" and "black" dualism, the study examined several groups residing in the same urban neighbourhood. The study demonstrates that in addition to the basic urge to live among "friends", individuals' decisions are affected by the relations to each of the other groups. In this sense, interflat, inter-building and intra-building segregation support each other to create and maintain residential patterns. Thus, individuals belonging to high-status groups are mainly sensitive to the fraction of people of their group. As assumed by Schelling (1978), those individuals are satisfied with a sufficient number of “friends” in a building and remain insensitive to the proximity of specific “others”. Conversely, individuals belonging to marginal and low-status groups are selectively sensitive to the presence of others: in addition to preferring buildings hosting a sufficient number of “friends”, they acquire micro-environments with low fractions of individuals belonging to inhospitable and high-status groups. A second contrast with Schelling’s model moves closer to the real world. There is a need to study multi-family buildings. Theoreticians avoid studying multi-parametric situations of this kind because of the multiplicity of “clues” demanded from the field. This study supplied these clues and managed to explore how the hierarchical position of each group affects the revealed behaviour of individuals.

The research has also revealed the simultaneous influence of building- and neighbourhood-level factors on the segregation and overcrowding pattern, which contribute to estimating the role of space in contemporary cities for retaining self-identities and reproducing social segregation. In this regards, Schnell and Benjamini’s (2005) “socially constituted spaces” concept is, actually, overlooked the interaction within residential buildings and local neighbourhoods that were found particularly relevant to religious and other cohesive communities. As the dwellers of the research area of Stamford Hill demonstrate, spatial patterns do reflect urban identities and important social relations in other urban communities and religious groups have active and meaningful parochial realms too.

Due to their high socio-religious status, many young Hasidic families who cannot afford to live inside the delineated area choose cheaper apartments, located in the margins, in buildings with other Haredi and non-Haredi households. Other families, all of whom belong to the Satmar move to Canvey Island. As we shall see in the next chapter, these families join the growing community as an extension of Stanford Hill. The clear demarcation line thus becomes a weak and flexible gateway, influenced by the penetration of other groups. The micro-segregation pattern could be relevant to non-religious situations, enabling strangers to live together in dense urban areas. Such micro-segregation dynamics are likely to emerge in the city hub, where relatively mixed dwellings host a wide variety of people. The micro-segregation dynamics operate at the level of the household, while the neighbourhood as a whole may appear to be socially mixed and highly integrated. While such a more in-depth investigation can reveal the typical footprint of the micro-segregation, further research may show the degree to which micro-segregation is a more general mechanism applicable to a broad spectrum of situations.

References

Alfasi N, Flint Ashery S, Benenson I (2013) Between the individual and the community: residential patterns of the Haredi population in Jerusalem. Int J Urban Reg Res 37(6):2152–2176

Anselin L (1995) Local indicators of spatial association—LISA. Geogr Anal 27(2):93–115

Bastian B, Loughnan S, Koval P (2011) Essentialist beliefs predict automatic motor-responses to social categories. Group Process Intergroup Relat 14(4):559–567
Baum A, Paulus PB (1987) Crowding. In: Handbook of environmental psychology, vol 1. pp 533–570
Beckett HE (1942) Population density and the heights of buildings. Trans Illum Eng Soc 7(7_IES-trans):75–80
Benenson I (2004) Agent-based modeling: from individual residential choice to urban residential dynamics. In: Spatially integrated social science: examples in best practice, vol 42, no 6–7. pp 67–95
Benenson I, Omer I, Hatna E (2002) Entity-based modeling of urban residential dynamics: the case of Yaffo, Tel Aviv. Environ Plann B Plann Des 29(4):491–512
Berman E (2009) Voices of outreach: the construction of identity and maintenance of social ties among Chabad-Lubavitch emissaries. J Sci Study Relig 48(1):69–85
Blumen O (2007) The gendered display of work: the midday scene in an ultra-orthodox street in Israel. Nashim J Jew Womens Stud Gend Issues 13:123–154
Borjas GJ (1998) To ghetto or not to ghetto: ethnicity and residential segregation. J Urban Econ 44(2):228–253
Breheny M (2001) Densities and sustainable cities: the UK experience. In: Cities for the new millennium, pp 39–51
Brown LA, Chung SY (2006) Spatial segregation, segregation indices and the geographical perspective. Popul Space Place 12(2):125–143
Bruch EE, Mare RD (2006) Neighborhood choice and neighborhood change. Am J Sociol 112(3):667–709
Cheng V (2009) Understanding density and high density. In: Designing high-density cities. Routledge, pp 37–51
Chizhov OP, Koryakin VS, Davidovich NV, Kanevskiy ZM, Zinger EM, Bazheva VY, Khmelevskoi IF (1968) Glaciation of Novaya Zemlya. Glaciol IX Sect IGY Program 18:338
Churchman A (1999) Disentangling the concept of density. J Plann Lit 13(4):389–411
Cimino R (2011) Neighborhoods, niches, and networks: the religious ecology of gentrification. City Community 10(2):157–181
Clark WA, Davies Withers S (1999) Changing jobs and changing houses: mobility outcomes of employment transitions. J Reg Sci 39(4):653–673
Dahan M (1999) The ultra-orthodox Jews and municipal authority, part 1–income distribution in Jerusalem. Hebr Jerus Inst Isr Stud Res Ser 79:1–50
Dougherty KD, Huyser KR (2008) Racially diverse congregations: organizational identity and the accommodation of differences. J Sci Study Relig 47(1):23–44
Dudley CS, Roozen DA (2001) Faith communities today: a report on religion in the United States today. Hartford Institute for Religion Research, Hartford Seminary
Elior R (1987) The concept of god in Hekhalot mysticism. Jerus Stud Jew Thought 6:13–64
Etkes E, Assaf D, Dan J (1999). Who believes not in the great leaders of all times. In: Etkes E, Assaf D, Dan J (eds) Studies of Hasidism. Hebrew University Press, Jerusalem, pp 125–149 [Hebrew]
Flache A, Hegselmann R (2001) Do irregular grids make a difference? Relaxing the spatial regularity assumption in cellular models of social dynamics. J Artif Soc Soc Simul 4(4)
Flint Ashery S (2015) Public welfare or sectarianism: a new challenge for planning. Plan Theory & Pract 16(3):299–318
Flint Ashery S (2018a) Schelling-type micro-segregation in a Hasidic enclave of Stamford-Hill. Hous Stud. https://doi.org/10.1080/02673037.2017.1414160
Flint Ashery S (2018b) Horizontal examination: micro-segregation mechanism in the diverse area of Whitechapel. In: Micro-residential dynamics, pp 25–34. Springer, Cham
Flint Ashery S (2018c) Effects of identity and belonging on residential experiences, perceptions and practices of the undocumented population in Whitechapel. In: Micro-residential dynamics, pp 73–94. Springer, Cham

Flint Ashery S, Alfasi N, Benenson I (2010) Micro-segregation of Haredim in Jerusalem. Israeli Sociol 12:81–110 (In Hebrew)
Flint Ashery S, Benenson I, Alfasi N, Bekman Y (2012) Schelling-like residential dynamics in a Haredi neighbourhood in Jerusalem. In: Diappi L (ed) Emergent phenomena in housing markets. Springer-Verlag, Berlin, Heidelberg
Flint S, Benenson I, Alfasi N (2012) Between Friends and strangers: micro-segregation in a Haredi neighborhood in Jerusalem. City Community 11(2):171–197
Fossett M (2006) Ethnic preferences, social distance dynamics, and residential segregation: theoretical explorations using simulation analysis. J Math Sociol 30(3–4):185–273
Freud D (2007) Reducing dependency, increasing opportunity: options for the future of welfare to work. DWP, London
Friedman M (1991) The Haredi ultra-orthodox society: sources trends and processes. The Jerusalem Institute for Israel Studies, Jerusalem (in Hebrew)
Giffinger R (1998) Segregation in Vienna: impacts of market barriers and rent regulations. Urban Stud 35(10):1791–1812
Gonen A (2006) Between Torah learning and wage earning: the London experience and lessons for Israel. The Floersheimer Institute for Policy Studies, Jerusalem (in Hebrew)
Gottdiener M (1997) The theming of America: dreams, visions, and commercial spaces. Westview Press
Green A (2001) Typologies of leadership and the Hasidic Zaddiq. In: Assaf D (ed) Zaddik and devotees: historical and social aspects of Hasidism. The Zalman Shazar Center for Jewish History, Jerusalem, pp 422–440 (in Hebrew)
Gropius W (1935) Theatrebau. Reale Academia d'Italia
Han Y, Taylor JE (2016) Simulating the inter-building effect on energy consumption from embedding phase change materials in building envelopes. Sustain Cities Soc 27:287–295
Holman C, Holman N (2002) Torah, worship and acts of loving kindness: baseline indicators for the Charedi community in Stamford Hill. Interlink Foundation
Iceland J (2004) Beyond black and white: metropolitan residential segregation in multi-ethnic America. Soc Sci Res 33(2):248–271
Ihlanfeldt KR, Scafidi B (2002) Black self-segregation as a cause of housing segregation: evidence from the multi-city study of urban inequality. J Urban Econ 51(2):366–390
Interlink (2010) The spatial needs of the Charedi community in Stamford Hill
Jacobs K, Manzi T (2014) Investigating the new landscapes of welfare: housing policy, politics and the emerging research agenda. Hous Theory Soc 31(2):213–227
Johnston R, Poulsen M, Forrest J (2007) The geography of ethnic residential segregation: a comparative study of five countries. Ann Assoc Am Geogr 97(4):713–738
Kasarda DJ (1978) Urbanization, community, and the metropolitan problem. In: Street D (ed) Handbook of contemporary urban life. Jossey-Bass, San Francisco, pp 27–57
Lipman VD (1962) The rise of Jewish suburbia. Transactions (Jewish Historical Society of England) 21:78–103.
Lofland HL (2009) The public realm: exploring the city's quintessential social territory. Aldine Transaction Publications, New Brunswick, NJ
Longley PA, Mesev V (2002) Measurement of density gradients and space-filling in urban systems. Pap Reg Sci 81(1):1–28
Lupo J (2004) Can Shas restore past glory. The Floersheimer Institute for Policy Studies, Jerusalem
Malpass P (2000) The discontinuous history of housing associations in England. Hous Stud 15(2):195–212
Martin L, Leslie M, March L (eds) (1972) Urban space and structures, no. 1. Cambridge University Press
Mills ES, Hamilton BW (1984) Urban economics scott. Foresman and Company, Glenview
Peach C (2006) Islam, ethnicity and South Asian religions in London 2001 census. Trans Inst Br Geogr 31(3):353–370

Portugali J, Benenson I, Omer I (1994) Sociospatial residential dynamics: stability and instability within a self-organizing city. Geogr Anal 26(4):321–340
Rapoport A (1975) Toward a redefinition of density. Environ Behav 7(2):133–158
Rey SJ, Folch DC (2011) Impact of spatial effects on income segregation indices. Comput Environ Urban Syst 35(6):431–441
Sakoda JM (1971) The checkerboard model of social interaction. J Math Sociol 1(1):119–132
Schelling TC (1971) Dynamic models of segregation. J Math Sociol 1(2):143–186
Schelling T (1974) On the ecology of micromotives. In: The corporate society. Palgrave, London, pp 19–64
Schelling T (1978) Micromotives and macrobehavior. WW Norton, New York
Schnell I, Benjamini Y (2005) Globalisation and the structure of urban social space: the lesson from Tel Aviv. Urban Stud 42(13):2489–2510
Sharma M (2012) A geographic perspective on intra-urban racial/ethnic diversity, segregation, and clustering in Knoxville, Tennessee: 1990–2000. Appl Geogr 32(2):310–323
Speare A, Goldstein S, Frey WH (1975) Residential mobility, migration, and metropolitan change. Ballinger Publishing Company
Stokols D (1972) On the distinction between density and crowding: some implications for future research. Psychol Rev 79(3):275
Tausch N, Hewstone M, Schmid K, Hughes J, Cairns E (2011) Extended contact effects as a function of closeness of relationship with ingroup contacts. Group Process Intergroup Relat 14(2):239–254
Telles EE (1995) Structural sources of socioeconomic segregation in Brazilian metropolitan areas. Am J Sociol 100(5):1199–1223
Waterman S, Kosmin B (1986) Mapping an unenumerated ethnic population: Jews in London. Ethn Racial Stud 9(4):484–501
Waterman S, Kosmin BA (1988) Residential patterns and processes: a study of Jews in three London boroughs. Trans Inst Br Geogr 79–95
Zhang T, Lin G (2007) A decomposition of Moran's I for clustering detection. Comput Stat Data Anal 51:6123–6137

Chapter 10
Canvey Island: The Effects of Group Action of Satmar Community

Abstract Since the early 1990s, a growing voluntary activity taking by interest groups interested in influencing urban planning policies has been noticed throughout the western cities. These groups create new channels for influencing actual planning. Special challenge creates non-liberal groups that use the rights and practice of planning to achieve sectoral goals. This chapter reveals the top-down strategy the Satmar community adopts to claim territory and influence the private and public space in the new settlement of Canvey Island. Purchasing the school made it possible to concentrate the community in a compact peripheral area at a relatively low cost. This collaboration is mostly explained by the attractiveness of Canvey Island which offers an affordable solution to housing problems within an hour drive from Stamford Hill. Finding such a resolution while relying on the community infrastructure increases the community's volume and the prestige of its leaders. Examining the role of different levels of cooperation in operating segregation processes, this chapter focuses on the abilities and limitations of the interest groups of Satmar to claim territory. In the urban reality, the inherent lack of collaborations and cohesion in the veteran local population limits their ability to safeguard their spatial rights and maintain their social practices. Today, Satmar is considered one of the most dominant congregations of the Torah world, and the veteran residents of Canvey Island are concerned about its impact on the public spaces, and its possible ramifications.

Keywords Group action · Residential segregation · Planning · Interest group · Institutional structure · Satmar · Canvey Island

10.1 Between Group Behaviour and Group Action in Canvey Island

The modern urban planning, influenced by local cultures and norms, the hierarchical mechanism was designed to guarantee the equal allocation of resources and appropriate standards of infrastructure for the various sectors (McKay 1996; Dean 2011). In urban reality, the tension between the worldviews of these different sectors and the system's difficulty in striking a balance between the various interests arouse a variety of political and spatial problems (Campbell 2002; Yiftachel 2006). The principal

S. Flint Ashery, *Spatial Behavior in Haredi Jewish Communities in Great Britain*, The Urban Book Series, https://doi.org/10.1007/978-3-030-25858-0_10

actors in the process—planners and politicians—had to cope with communities characterised by diverse institutional structures, access to resources, spatial behaviour, inconsistent interests, and various other types of discourse and concept (Taylor 1998; Stone 2002; Flint Ashery et al. 2014; Flint Ashery 2017).

In social discourse, it is common to distinguish between moral stance, political philosophy, ideology and social concepts of individualism that emphasise independent judgment and desires, valuing independence, self-reliance and collectivism that emphasise interdependence and cohesion within social groups (Sack 1986; Saegert and Winkel 1990). Whether collectivism is horizontal, i.e. stresses collective decision-making among equals, or vertical (Flint Ashery 2018a), i.e. based on hierarchical structures of power and on moral and cultural conformity, collectivists are individuals who view themselves primarily as parts of a whole, be it a family, tribe or nation (Paasi 1996; Nylund 2001). Although individualism and collectivism are seen as opposites, they are connected, and the concept of individualism is often incorporated in the idea of collectivism (Newman 2003). Encounters between the two concepts occur in the modern city, where the social and cultural aspects of the populations are reflected in their different lifestyles (Johnston et al. 2007). As opposed to indirect collaboration, the asymmetrical balance of power between neighbouring communities creates group action (Boal 2008). When the group members realise they are more likely to achieve their goal when acting coordinately and simultaneously, they may consolidate under a clear authority and use their enclave as a base for group action. The active aspects of this behaviour, previously described in conflictual cities, such as Belfast, Nicosia or Sarajevo, the authority will use its access to capital—whether financial or knowledge-based—to claim the territory for themselves and expand it by forcing out everybody they deem not belonging (Gottman 1973; Malmberg 1980; Sack 1983). Practical, ideological and other considerations affect the spatial behaviour of users, challenging liberal planning systems (Altvater 1998; Flint Ashery 2015, 2017), particularly those of groups that are committed to other laws and concepts, like the Haredi groups.

"Interests" are referred to or defined as motivations for action (Oppenheim 1975). While private interest represents the values and beliefs of western individualistic culture (Miller 1999), the public interest is conceived as a value commitment to benefit the collective (Funk 2000). Both affect our judgment and choices (Kong 2001). Private interest can be assigned to groups and collectives and not only to individuals. These specific interest groups, characterised by a well-defined organisation and constituencies, have better access to capital, whether financial or knowledge-based, which they deploy on behalf of particular goals (Salisbury 1969; Denzau and Munger 1986; Beyers 2004; Hendricks 2006). Interests are considered as efficient predictors of human behaviour (Hechter and Kanazawa 1997). People are brought up believing that self-interest will promote self-benefit so that private interest motivates their social behaviour and actions (Ratner and Miller 2001). Hence, without vested interests, people refrain from acting or expressing opinions about an issue, believing that since the issue is of no personal concern, it is not their place to react (Miller 1999).

Certain norms may lead an individual to be inclined towards public interest. An application of rational choice theory—the "thick model", is concerned with communal outcomes, examining norms and other social institutions to envisage possible modes of action (Ratner and Miller 2001). Other models describe and predict altruistic behaviour, in which normative values and social factors motivate individuals to act unselfishly (Andreoni 1995). Funk (2000) demonstrates that individual voting patterns are guided by social concerns, i.e. the setting aside of private interests for the benefit of society (Kolossov 2005; Boal 2008). When policy topics with public visibility arise, interest organisations such as corporations and/or government agencies participate in citizen forums and are inclined to express public interest motives (Hendricks 2006). The outcome will be influenced by the diversity of the interest organisations involved, especially by the balance between profit and non-profit interests affecting public perception of policy (Lowery et al. 2005).

Although the institutional structure of the relevant groups, their organisation, access to government authorities and sources of funding allow us to analyse them according to their type of interest group (Drezon-Tepler 1990; Sharkansky 1996), these terms are not absolute, and therefore, this research will not make use of them but concentrate on the group's characteristics. The modern society is composed of a multitude of subcommunities with inconsistent interests. Such a condition renders a government's effort to acknowledge the public interest quite difficult (Clarck 2000; Poppelaars 2007). This difficulty is further intensified with the average citizen's tendency to refrain from public policy involvement, which does not necessarily represent a true general public interest.

Most researchers attribute these processes to social, economic and political factors rather than to group behaviour in urban space (Aran et al. 2008). Hence, there is a conspicuous dearth of research dealing with the effects of group behaviour on the design of residential areas. To examine these processes, one must turn to the social systems that drive them and their values. Although the Haredi ideology and its collectivist structure are often portrayed as fundamentalist, guided by authoritarian power (Almond et al. 1995), this chapter discusses the collective features of fundamentalism and their urban expression. Likewise, although many researchers attribute the gap between top-down statutory planning and bottom-up dynamics of planning to the gap between the certainty-oriented regulatory approach and the flexible dynamics of planning; the planners being either professional or political (Cullingworth 1994; Booth 1996; Tewdwr-Jones 1999) or even to the concept that mockery of the legal system by leaders and citizens alike vitiates the spirit if not the letter of the law (Sprinzak 1986; Alexander 2002; Feitelson 2011). This research, however, explains this gap as the outcome of the liberal planning system's inability to cope with groups using liberal practices to achieve sectoral goals.

This chapter deals with a non-violent group action of the Satmar community in Canvey Island. The literature assumes that a collective that gathers at a given place and time creates direct cooperation and is defined as group behaviour (Saegert and Winkel 1990; Anderson and O'Dowd 1999). Clear authority reaches consensus, sets rules and creates an organised segregated pattern. The group emphasises its organised power that differentiates it from the surrounded non-organised individuals. When group members realise they are more likely to achieve their goal when acting coordinately and simultaneously, they may extend their enclave through group action (Granovetter 1978; Kong 2001). This behaviour, characteristic of groups such as group purchasing organisations and gated communities, is reinforced in traditional communities where religion acts as social cement (Bankston and Zhou 1995). In these societies, similar interests and non-territorial norms of control are completed by territoriality that creates a stronger socio-spatial order than is usually found in communities based on the needs of the individuals (Anderson and O'Dowd 1999; Kolossov 2005). Today, Satmar is considered one of the most important congregations of the Torah world. Satmar is seeking a housing solution for the growing community of Stamford Hill, and the veteran residents of Canvey Island are concerned about its impact on the public spaces, and its possible ramifications.

10.2 The Research Area of Canvey Island

Canvey Island (see Fig. 10.1) is a civil parish and reclaimed island in the Thames Estuary in Essex, England. It has an area of 18.44 km^2. Canvey Island is separated from the mainland of south Essex by a network of creeks. Canvey Island, lying just above sea level, is prone to flooding at exceptional tides but has nevertheless been inhabited since the Roman invasion of Britain. In 2015, the population of Canvey Island comprised about 37,500, inhabiting 10,863 houses and residential buildings. Initially, most of the apartments in Canvey Island were about the same size—68–74 m^2 (2 bedrooms), though some of them were enlarged. Canvey Island is populated mainly by the white British population, along with small Asian and Black communities. Seventy-four per cent of Canvey Island's inhabitants identify themselves as Christian (including CofE, Catholic, Protestant and other denominations), 0.2% describe themselves as Muslims, 0.1% as Hindus and more than 16% describe themselves as having no religion.

The Jewish Congregation of Canvey Island (JCoCI), a Jewish congregation of Ashkenazi orthodox, was founded on Canvey Island in 1935 as an offspring of Southend community but closed by 1938. In 2016, Haredi families from the Satmar Hasidic group gets priced out of the Stamford Hill area of North London and have been establishing a new community near the former Castle View School, in the Northern region of Canvey Island town centre (see Fig. 10.2).

10.3 Planning Agenda

Protocols of the Development Control Committee (7 February 2017 and 7 March 2017) indicate that the Local Planning Authority has acted positively and proactively in determining application number 16/0917/FUL by identifying matters of concern within the application.

According to the protocols, the applicants seek consent for the change of use of the building from D1 (non-residential institution) to C2 (residential institution) and to form ten dormitories, accommodating some 60 students, with staff in attendance,

Fig. 10.1 Canvey Island, in the Thames Estuary in Essex

at first floor level within the existing building. The applicants have advised that the school currently has 54 students ranging in age from 17 to 20 and those students are not permitted to use/have a car at the school. The applicant further advises that the School would normally have about six teachers. The provision is associated with the lawful use of that building as a school.

The Development Control Committee see the proposal as also represent a highly sustainable form of development. The application site is located on land allocated

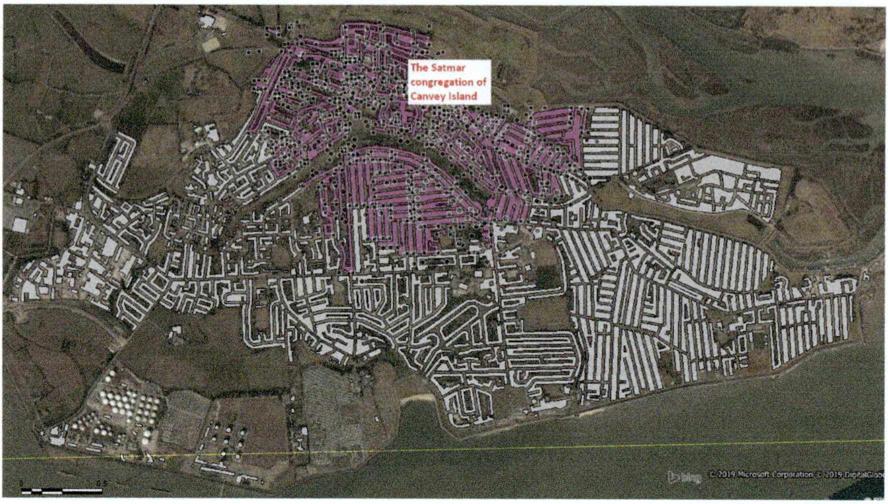

Fig. 10.2 Headquarter of Satmar congregation of Canvey Island. Preplanning was evident in the acquisition of properties located within a 20-min walk from the centre

for Green Belt and School purposes within the adopted Local Plan. Paragraph 90 of the NPPF confirms that the reuse of such buildings is not inappropriate development in the Green Belt provided that the proposal preserves the openness of the Green Belt and does not conflict with the purposes of including land in Green Belt and in the case of the reuse of existing buildings that the building to be used is of permanent and substantial construction. Paragraph 70 of the NPPF states that the planning system should guard against the unnecessary loss of valued facilities and services, particularly where this would reduce the community's ability to meet its day-to-day needs. Policy SD1 of the New Local Plan states that when considering development proposals, the council will take a positive approach that reflects the presumption in favour of sustainable development contained in the National Planning Policy Framework. As stated above the provision of dormitory accommodation seeks to support the reuse of an existing school building which in itself represents a highly sustainable form of development. The provision of dormitory accommodation further supports this level of sustainability by limiting the necessity for traffic movements to and from the site and creating a resident base which would contribute to the local economy. The proposed dormitory provision is therefore considered to be highly sustainable in the context of the continuing use of the school building for educational purposes.

The application specifically seeks to provide dormitory accommodation to serve a school, in a building previously used for mainstream educational purposes—a role which it is no longer required to fill. The reuse of the building as proposed would make good use of an existing building and would represent a highly sustainable response to the current redundancy of the building. Paragraph 70 of the NPPF states that Planning Authorities should ensure that established shops, facilities and services are able to develop and modernise in a way that is sustainable, and retained for the benefit of the community, and paragraph 72 of the NPPF identifies that the government attaches great importance to ensuring that a sufficient choice of school places is available to meet the needs of existing and new communities. Policy HC11 of the New Local Plan states that the council will support proposals for new, altered, extended or replacement education facilities in order to allow communities to meet their day-to-day needs, where it can be demonstrated that the development will respond to the modern needs of the local community and provide flexible space that can respond to the changing and specific needs of the local community and where appropriate be capable of offering safe refuge in the event of a natural or manmade incident occurring. Furthermore, the reuse of the school as a faith school, as is suggested, would add to the diversity of educational opportunities present within the Borough while the provision of dormitory accommodation would assist in meeting the day-to-day needs of the growing Jewish community. As a result, the Local Planning Authority has been able to grant planning permission for an acceptable proposal, in accordance with the presumption in favour of sustainable development, as set out within the National Planning Policy Framework.

10.4 Methodology: The Case Study of Canvey Island

To investigate the effect of a group action on territoriality and reshaping of the neighbourhood, the study constructed a detailed spatial-temporal database that contains exact geo-referenced data on Canvey Island newcomers. The first step was to identify the Hasidic territory, within 20-min walking distance from the school, as orthodox Jewish communities need to live within a short distance of a synagogue to enable them to obey the laws of the Sabbath, which prohibit driving or riding in a car or any other vehicle from sundown on Friday until sundown on Saturday. The second step was to identify the purchases that took place in this area since March 2016. In each house, the residents were asked about the identity of the residents of the building and whether they rented or owned their apartment. Although cooperation was limited, most of the new residents (53 families) indicated that they belong to the Hasidic community of Satmar, based in Stamford Hill, and gave the date of entry into the new apartment. To reconstruct the dynamics of population replacement, 46 veteran residents—who sold their apartments between 2016 and 2017—were identified and interviewed about the sales process. It should be noted that most of the people who left were living outside the island, and the interviews were conducted by a telephone conversation. They provided information about the price (or the price in relation to the market price) and the month/year of the sale. Ex-owners were willing to reply to these questions in 100% of the cases.

To complete the fundamental part of this research, 32 interviews had conducted with key figures such as community leaders, Canvey Island's Mayor, municipal planners, real estate agents and people from the public. Interviewees were chosen on the basis that they offer a range of different types of knowledge and perspectives on their community. Among the interviewees also reside the veteran inhabitants who could provide explanations and describe the processes taking place in the neighbourhood from their point of view. The cross-referenced data produce information on the population exchange and express the dynamic processes. The interviews were also assisted in the identification of further vital contacts.

Construction of the Canvey Island GIS was based on the layers updated for the year 2018, which were provided by the ordnancesurvey.co.uk/opendatadownload/products.html. The characteristics of all the research area's apartments and households were organised as a GIS layer, in which every record in the table is related to the corresponding building. Since most of the buildings are private houses (Flint Ashery 2018b), mainly typical British detached bungalow, and the relatively small number of the Hasidic occupant, the collected data were analysed but will not present on a map to protect the confidentiality and keep the building-scale resolution. The layer was then included in the area's high-resolution GIS. Canvey Island GIS contains additional layers pertaining to topography, roads, land parcels and buildings, the latter characterised by use and number of floors.

10.5 Top-Down Initiations: The Hasidic Community of Satmar

Satmar is a Hasidic group originating from the city of Szatmárnémeti, Hungary, where it was founded in 1905 by Rabbi Joel Teitelbaum. With about 70 K people affiliated to the group, Satmar is one of the largest Hasidic dynasties in the world. Satmar is characterised by extreme religious rigidity and rejection of modern culture and also serves as the leading power within the Haredi circles which oppose the State of Israel. To reinforce opposition to secular studies, Satmar sponsors a comprehensive education and media system in Yiddish, and its adherents use Yiddish as a primary language. Satmar is the largest congregation in Stamford Hill with six directly associated synagogues.

In recent years, demographic, social and economic pressures have limited involvement of the leadership in the Stamford Hill community's daily life, and its ability to enforce group discipline for continued residence within the community has been weakened. By default, the leftovers are driven to establish residence far from the original group enclave, and scouting parties were sent out to Canvey Island, Milton Keynes, Harlow and Westcliff, to check affordability and suitability to the community needs. The geographic separation that is created between the generations threatens the continuity of the community: on the one hand, the population living in the original enclave is ageing, and on the other, the lifestyle in the new communities tends to adapt itself to the new conditions of life and sometimes adopt a more extreme lifestyle. Interviews with key figures in the community, such as Rabbi Abraham Pinter, the spokesman for the orthodox Jewish community (20 December 2017), Isaac (pseudonym, male, 31), who runs the neighbourhood patrol (20 December 2017), Josh (pseudonym, male, 32), one of Canvey's pioneers (20 December 2017), revealed that these processes which were a source of concern for the leaders of the Satmar community from Stamford Hill, motivated Rabbis Wosner, Pollack and Stolzberg to encourage group action to expand the origin living space of the group.

Rabbi Pinter describes: "There's a family here, all lived together at the parents' four-bedroom house. The parents with their six children—two of them are married and also lived with their families in the adjacent rooms. It was very crowded, and the young children could not concentrate on their studies. It was always very noisy, and everyone was nervous. The window was broken, and there was no money to fix it, so they used a plastic bag. It is, of course, a very typical case. So five years ago we started to look for a solution. We realised that the community could not stand these housing prices. It is not only for the poor, not just the middle classes, but a general, overall, huge problem. Even the wealthiest cannot afford to buy houses for their children. And if you can't buy a house for your daughter, she won't be able to marry with 'Ben Torah' (Torah scholar)".

According to the founder of Interlink Foundation, the membership organisation for 200 orthodox Jewish charities, champion of the needs of Britain's growing orthodox Jewish community by advocating for community needs and helping influence policy makers to work equitably and collaboratively with the community is also aware to the limitations of the council: "We went to the Hackney council as we have some connections there. They tried to help, but they were limited. The solution proposed by the council is offering us one apartment here, and one apartment there. Remote from the community or in inappropriate locations. Not something that could be appropriate for a community that grew by a thousand families a year and had to live together". However, despite the involvement of Stamford Hill's Council, the official solutions are inadequate for meeting the demand. As a result, Hasidic pressure on the enclave has increased.

Individuals' identification with their community and their desire to raise their children in a similar religious environment evoke conscious ideological and moral recruitment (Bankston and Zhou 1995; Chakak 2004). In addition to the threat of non-acceptance to the group's institutions, clear-cut monitory mechanisms ensure compliant behaviour, allowing the leadership to impose discipline and organise matters following its preferences (Hasson 2001; Gonen 2006). The expressive incentive for the groups' member is that they would help out in the process of trying to obtain the group's goals (Valins 2003). The chief executive of the "Agudas Israel Housing Association" (AIHA) asserts: "House prices of £1.2 million-plus in Stamford Hill are beyond the reach of many families. I am now encouraging families who desperately need good quality, affordable housing to relocate to Canvey—something I wouldn't have thought possible 18 months ago. However, Hackney Council is 'not interested' in helping the Haredi community to expand locally, it is not supporting our housing needs, and grants for affordable housing from the Greater London Authority were worth a fraction of what they used to be. Buying properties in Canvey Island to rent out adds the option of affordable rental housing for people wishing to move to the area who are not in a position to buy a home at the moment" (20 December 2017). The Satmar group is characterised by a well-coordinated institutional structure, organised funding, and members' strong commitment; these characteristics enable them to mobilise dedicated manpower and significant resources to realise the community's segregated preferences.

10.6 Evaluation of Canvey Island's Space–Time Dynamics

To understand how the group action works, this research examined the penetration of Satmar to Canvey Island since May 2016. The rapid process indicates the abilities of the organised community when it competes with non-organised individuals. The first step was critical to the success of the "Haredification".

Community's representatives identify suitable community centre/school for sale, and a benefactor has bought Castle View School on the island for the Hasidic pioneers on 24 March 2016 from Essex County Council.

1. Six lone pioneer families identified apartments for sale and moved in over the same weekend in May 2016.
2. Massive penetration turns "non-hostile" into "friendly" area with the settlement of forty families (2016–2017).
3. The rapid movement patterns were well-planned. The designated area was "marked" by the leaders of the community. Preplanning was evident in the acquisition of properties within a 20-min walk from the Jewish Congregation of Canvey Island (JCoCI, the former Castle View School), see Fig. 10.2.

Aaron (pseudonym, male, 28), one of the pioneers explains the effect of this top-down initiation on his life: "Like any other Haredi community, the community of Canvey Island needs its community-oriented organisations. We sought for donation, and a family of philanthropists paid £1.75 million GBP for the former school building and its attached playing fields. In addition, this new school will have residential dormitory accommodation for around 60 boarder students on the first floor of the building. Returning the school to an educational establishment is also good for the locals, as the school has stood empty since 2011. Today, we are already a community, and the boys are studying here at our private school, the one we have established. The girls, however, still need to travel every day by public transport to Stamford Hill. This is only until we will have enough girls to open new classes .... Being a pioneer here means a new start here, as you cannot work at your previous job. So now I am teaching at the new school, devote time to building our community and maintaining good relations with our neighbours—both for ourselves and for those who will come. Yes, we are an integral part of our original community. In the beginning, my wife did not want to move. It is isolated and also below sea level—dangerous waters. But now, we are renting house owned by the community and we have a much better education here, with small classes. The actual idea of front garden space, detached, we do not see it in this on Stanford hill. It is a change and a place that will make them get used to a different kind of people. A fresh start, loads of opportunities for us".

The head of JCOCI Educational Foundation and one of the pioneers describe the second stage of the process: "AIHA commits innovative seeking solutions to

Stamford Hill's housing crisis, community support for the next generation. Being at the forefront of housing for decades, AIHA acutely recognises the challenge facing the Haredi community, and by purchasing homes in Canvey Island, it is doing a great deal in 'making it happen. As a community, we were able to close a deal for a private school and community centre and enlarge the community's area. When the families cooperated, we were able to buy several houses in the area. We are buying through the local agents, several properties together, so we can better negotiate and keep the price down, and we are also knocking on doors, offering over the price. It is true that in today's market someone may be able to find a cheaper property. But buying outside of the group means that we won't accept his children to the educational institutions of the community. This is our way to impose community discipline, and it works, as we seen also for the Litvish in Westcliff and Gateshead. In the beginning, we were six families, and everyday men came from Stamford Hill to complete a Minyan (Ten Jewish adults required for certain religious obligations). They did not leave us alone, brought us Kosher food. There are only Satmars and the non-Jewish population here, and there's nothing to compare: the children are blooming. With two or three children in a room, they can study adequately. There are a garden and clean air. The children are proud of their new home".

Lyva (pseudonym, female, 28), another one of the pioneers, describes the challenges: "I think the biggest challenge for people is to take the plunge and move out of Stamford Hill. Currently, we have not much to offer. We've got a long way to go to build this community. When you move a community over, you have to have the infrastructure, the synagogues, the schools, the shops and everything a community needs. We've got a little tuck shop here for the school, but it's very, very basic. It resonates in that way to the shtetl on eighteenth century Poland, and these little core villages with a little synagogue and a little shop, and the tinker, the tailor, the candlestick maker, that kind of old Jewish model. To encourage more people from the community to move to Canvey Island, the community have bought a house for the families to try out the island for the weekend. We're aiming for 60 families within a year or so". In order to understand how group action works, the next section examines the rapid occupation of the area around the former Castle View School by Satmar between 2016 and 2018.

Table 10.1 compares the quality of life (HQI) of Stamford Hill's and Canvey Islands' residents, following the in-depth survey. According to the Housing quality indicators, the HQI level in Canvey Island is more than twice higher than in Stamford Hill. Although most of the residents appreciate the location of Stamford Hill, mainly due to the services it offers and the proximity to institutions, both Canvey Island's newcomers and Stamford Hill's residents who visit in Canvey Island can appreciate the quality of life in this new settlement.

Table 10.1 HQI assessments in Stamford Hill and Canvey Island according to the housing quality indicators (Homes and Communities Agency, 17 May 2011)

	Stamford Hill	Canvey Island
Location	10	4
Site—visual impact, layout and landscaping	2	10
Site—open space	4	10
Site—routes and movement	4	4
Unit—size	0	10
Unit—layout	0	8
Unit—noise, light, services and adaptability	4	10
Unit—accessibility within the unit	2	10
Unit—sustainability	5	5
External environment—building for life	3	10
Combined score	34	81

The data were collected directly from the residents of the two neighbourhoods

10.7 Avoiding the Emergence of an Internal Housing Market

The Hasidic landlords are busy buying houses. Data obtained from local estate agents (Quirk, A. 14 December 2017; Fitzsimons, C. 14 December 2017; Poyntz, R. 14 December 2017; Lane, S. 17 December 2017) and former residents enabled reconstruction of the gradient of prices and to examine the market dynamics as of 2016 (Fig. 10.3):

1. Penetration: four apartments purchased by Satmars at the market price.
2. Before the "tie-break": apartments 5–7 purchased above market price.
3. Avoiding the emergence of an internal market: due to the unique appearance of the Haredi, the area surrounding JCOCI becomes identified with the Satmar community. The area, as a whole, becomes more expensive and creates property line around/within its boundaries.

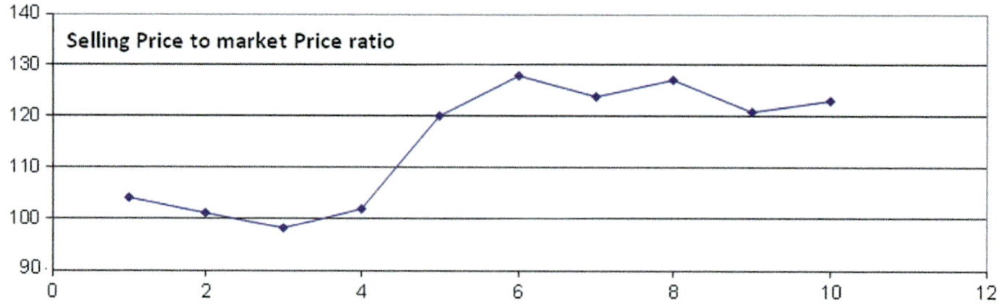

Fig. 10.3 Average selling price to market price ratio as dependent on the sequential number of apartment among those sold by the veteran to Satmar, for each postcode area (2016–2018)

The realtor Richard Poyntz describes how processes within the Satmar community affected prices: "The community organisation secured fixed housing price for a limited time. Everything is centred around the institute of the Jewish Congregation of Canvey Island (JCoCI). The original deal was for it to be a complex of about 50 flats, but all of a sudden a private school has come along, and as I understand it they have offered more money than the profit from the flats would have been". In Zoopla's Heat Maps, the area appears as more "reddish" and indicates the level of attractiveness of the region.

The Haredi wish to avoid the emergence of an internal housing market in Canvey Island was explained by Director of Housing at Industrial Dwellings Society (IDS), a housing association that specialises in providing housing for the Jewish community, that manages around 1500 properties in Hackney, Tower Hamlets, Southwark, Redbridge, Haringey, Camden, Barnet, Hertsmere and Canvey Island: "There are 40–50 different sub-communities in Stamford Hill, probably even more than that. Each community is a member of the Union of orthodox Hebrew Congregations. Each community has its independence, ideology and infrastructure, and often will compete with other communities over a suitable property, and there's no way we can attempt to control the inner market there. One example for inner-competition was over the former Skinners School grounds. Everyone was competing to buy this off Berkeley Homes after the council turned down an application to turn it into homes at the end of 2010. The Belz and Lubavitch groups joined forces to buy the site and believed that Satmar had lost interest. But then a third party (The Satmar group) outbid the others in a one-night negotiation that prevented the others from bringing higher offers to the table—(Jewish Chronicle reported the deal). To someone outside of the community may be the growth looks organised and aimed at general expansion of the community when in fact it is driven by internal competition. However, by selling directly to the families, we can avoid the emergence of an internal housing market. People are not in fear that apartments will be 'snapped up', and that they will lose the opportunity to live among friends, there is enough space. Belz community went to Westcliff-on-Sea, and Satmar came here. However, everyone wants to live in a walking distance from the school. This causes housing prices in this specific area to rise a bit. The market which drives the price rise here is not internal, but mixed".

The Chief Executive at IDS explained that a former member of the IDS board had initially been made the suggestion that the organisation become involved in the Canvey Island project: "In Canvey Island, a new community is being established, and IDS was keen to help in addressing the severe housing problems and overcrowding previously experienced by the new residents. We had investigated in the enterprise and bought four new-build properties in the area, designated as "affordable rent" properties. There are two four-bedroom houses and two three-bedroom houses. We are hoping to purchase a fifth property in due course, as part of a parallel arrangement with AIHA. The properties are offered to the firsts on the list, the neediest families, who meet the criteria".

As of March 2016, Haredi direct collaborations have succeeded in strengthening the Satmar Haredi group's presence on Canvey Island. According to the Chair of Interim Executive Committee Residents/Workers Representative "in a diverse urban situation, a small group of well-organised people with the backing of any one section of the community can impose their views on a very wide area". Although group action motivated by distress and hardship, as well as by ideological practices expands the group's territory, the sporadic purchases are integrating the group within the veteran population of the neighbourhood and create a heterogeneous pattern.

One of the local veteran residents who welcome the new Satmar settlers explain: "Canvey Island has lost his youngsters. Schools were closed, and it became a sad place to live. I know that these people are different from us, but historically Canvey Island has a good relationship with its Jewish community. Although this community is more religious, I hope that they will also integrate and give a blessing to this place".

10.8 The Hosting Community

One of the five most pro-Brexit wards in Britain, Canvey Island was voted the most English place in Britain in 2013, with 73% voted to leave the EU. The Deputy Town Mayor of Canvey Island Independent Party explains: "When they did the last census, most people who live on Canvey, instead of putting British, they put English. It's just that they felt that they were English more than anything else … We have heard about them (the Hasidic community) moving to Canvey. Obviously, they've got their own traditions and their own religion and that lot, so that will probably stop them mixing as much as what someone else would. I know that obviously, both communities have got to get used to each other and all that lot, but how much will their religion allow them to be part of the community? Because there are obviously certain things they can't do on certain days and things like that … In a way, I think that they choose to distance themselves. I'm not saying I want them to integrate with us totally, I'm just saying that I think they won't be able to integrate as much, because of the rules of their religion" (BBC, 9 January 2018).

The direct cooperation of the Satmars creates pressure on the residents. These residents comprise individuals with diverse beliefs and interests. Most of them are autonomous homeowners. Confronting the new incomers who are backed by capital and group support, the veteran residents have no community support. Table 10.2 indicates that the level of exposure of the two veteran groups to the Satmar community is relatively similar. However, the level of familiarity with the Satmar community is relatively low, especially among those who have sold their properties and left the neighbourhood. The residents who live next to the newcomers are very interested—double than those who left—in getting to know the Satmar community and are willing to contribute that the Satmar will integrate into the general society of Canvey Island. The veteran residents, and especially those who left the area, are also interested that the Satmar will integrate into Canvey's society and believe that it is possible to a large extent despite their fear of the impact of this process on the existing population.

Retired gas fitter Bill (pseudonym, male, 71) is a hardened Brexiteer. He helps run Canvey Island Rugby Club, and his house located at the back of the new Hasidic school and he is curious about his new neighbours: "I know they're tribal … They

Table 10.2 Declared attitudes of the veteran residents of Canvey Island: those who sold their properties to the newcomers of the Satmar community and those who live nearby

Stated	Attitudes of those who sold their properties to Satmar	Attitudes of those who live nearby
$N = 46$	$N = 46$	$N = 32$
What is your level of familiarity with this community?	1.2	3.7
What is your level of exposure to this community?	2.1	2.3
How much do you want to get to know this community?	2.3	4.6
How much do you want the community to integrate into Canvey's society?	5.0	4.7
To what extent do you think the community can integrate into Canvey's society?	3.7	3.1
How much do you want and willing to contribute that this community will integrate into Canvey's society?	2.3	5.0
How concerned are you about the level of integration and its effects on this community?	3.5	3.2

In average, ranked from 1 (lowest) to 5 (highest)

keep themselves to themselves and they wear funny clothes. I've got no problem with them. As you can see, we've got the English flag flying over there, so they'll know where they're living. At the moment, let's be honest about it, at the moment". Other neighbours, Gary and Sharon (pseudonym), wish to learn more about the Hasidic community: "I like to know why they've moved on to the island, what's brought them to the island … I want to know what their daily routine is. Yeah, I would like to know that because I'd be able to tell them mine, so I'd see if they were on par with mine! Do they go to work? I don't know. Do they pray all day? So they need to be trusting the Canvey people, and I think coming out a little bit more and understanding us so we can understand them" (BBC, 9 January 2018).

Reverend Tudor, the leader of St Nicholas' Church wants his congregation to welcome the new arrivals: "… and they are searching for a better understanding of God, and they do it by separating themselves from the world. By focusing on their Scriptures, the stories of their ancestors, the rules and regulations that were passed down through the ages. They focus on what to wear, what days to work, and the children are allowed mobile phones … And "although the books that we call holy have many differences, the one thing in common that our books have is the welcome of the stranger. And so, in a sense, we're all strangers, we're all travelling through this earth, and in a sense, we all belong, and we're the same family" (BBC, 9 January 2018). Speaking at a Chanukah event hosted by the Hasidic residents at Canvey Island Reverend Tudor states: "Whatever our religion, whatever our tradition, what we have in common is our humanity" (14 December 2017). This attitude is highly supported by Chris Fenwick, a lifelong Canvey Islander, pub landlord of The Oyster-fleet and manager of the rock and roll band Dr Feelgood, who initiates meetings and tours to integrate these two unlikely communities (14 December 2017).

The chief executive of the "Agudas Israel Housing Association" (AIHA) talks about the chances to bring these two communities together: "Look, I have never experienced a non-Jewish community so welcoming, so open-hearted and so friendly to us Jewish people. But our own very strict rules and regulations, and our tolerance … Because of Jewish history, I am worried about prejudice. That's always in the back of people's mind. And there are things that we do face. People see us for the first time, we expect they will ask questions, and some will rather shout something derogatory. I hope that we will not let them down. That we will live side by side, we let them do what they want, and we do our lifestyle" (14 December 2017). One of six initial families explains that "JCoCI plans to refurbish a former bank on Canvey High Street to create a kosher bakery … If it goes through, it is highly likely many of the customers will not be Jewish … Serving daily fresh bread with things like the sweet bread challah is something we don't have a lot of here, hopefully, if all goes well we can make it happen." (Jewish News 2019).

10.9 Public Welfare or Sectarianism?

The decision of some dozen Haredi Jewish families of Satmar to move to Canvey Island, more than one-hour drive away from Stamford Hill, where their families have lived for several decades, derives mainly from the soaring house prices. Haredi Jews often have large families, with up to seven or more children, means that they need large properties, both for the current needs, and both for their children, that, in turn, will have large families of their own. Therefore, the need for many residential solutions in the same area is of great demand. For the Haredi, scripture and its interpretations are the source of not only religious, but also social, economic and political authority, dictate their daily life. Haredi individuals cannot choose their residences on a personal basis, as they require communal facilities, such as a particular educational system, a synagogue and Kosher shops. Their communities are very self-contained within their enclave, means that any residential move has to be a group move. The decision to select Canvey Island as a venue for establishing an offshoot of the community makes sense as this location is not too far away from their existing base, so the newcomers to Canvey can rely on the community and its institution, Kosher shops and Gemach. Also, the house prices in this scenic stretch of the Essex coast are low. Other Haredi communities settled nearby, in Westcliff-on-Sea and Southend-on-Sea, about half-an-hour drive from Canvey Island.

Generally, Canvey Island almost exclusively populated by white population and is known as one of the most pro-Brexit parts of the UK. As part of the greater liberal planning system, Canvey's municipal authorities respond according to the premise that all residents should receive services that meet their lifestyle needs. However, limited resources entail regulation, which is influenced by power relations between the various population groups. The Satmar community is aware that their rapid entry into Canvey Island may overload the infrastructure and independently purchased the compound, which is supposed to contain the synagogue, the educational institutions and the boarding school. These movements indicate that the community have a duty towards God and public resources are necessary to help them fulfil their obligations, and there is nothing to negotiate. Nonetheless, they can dissociate themselves from Torah to reach their worldly goals.

The first signs from the local residents have been positive, as the veteran residents were pleased to welcome some new families to their area which attractiveness has deteriorated over the past years. Rabbi Jonathan Romain (2018) describes that "Those who have already interacted with their new Strictly—orthodox peers have admitted that they find their lifestyle strange, but also have praised them for being good neighbours. They are a remarkably law-abiding group, with hardly any instances of public disturbances, drunken parties or drug problems". However, it is anticipating that other families will join the families who are already present on Canvey Island and that the Satmar community will expand. Similar to other Haredi communities, such as Manchester's Satmar community and their home community in Stamford Hill, they will treat the others with courtesy and civility, while also become entirely isolated. Although the influence of the veteran residence on the planning policy correspond previous articles argue that such an established community can insert public interest into the policy process, their potential to influence the planning policy is indirect and partial, while the influence of representatives of Satmars is direct and effective. Thus, the limited ability of unorganised groups to protect their interests from being taken over in the dynamic and pluralistic city presents a new challenge for planning.

10.10 Discussion: Group Action as a Driving Force for Neighbourhood Change

This research has revealed the top-down strategy the Satmar community adopts to claim territory, distinguish their abilities and limitations to influence the neighbourhoods' private and public space. The in-depth survey indicates the way direct, effective collaborations of individuals of the Satmar community with their leadership led to the new settlement in Canvey Island. The unique values and lifestyles of the Satmar Haredi community and the general society are reflected in their struggle to occupy space and claim resources, both between the groups and in their dealings with institutional planning systems. Purchasing the school made it possible to concentrate the community in a compact peripheral area at a relatively low cost. This collaboration is largely explained by the attractiveness of Canvey Island which offers an affordable solution to housing problems within an hour drive from the traditional enclave of Stamford Hill. Analysis of collectivism versus individualism in a liberal competitive economy reveals important elements in the ability of a cohesive community with high demographic fertility to harness private interests and to wrest control of space from other groups, thereby creating a "contiguous" Haredi living space. The Satmar group has deep social and cultural roots and strong mutual interests. The organisation of the group allows communal norms and social ties to further private interests. The group's collective and disciplined activities and territorial behaviour conform to its leaders' spatial preferences.

This behaviour, spreading to the peripheral streets of the neighbourhood demonstrated that the Satmars are effective strategists in planning, organising and executing a heterogeneous space that is characterised by its laws, although being also accepted by the veteran locals. In the urban reality, the inherent lack of collaborations and cohesion in the veteran local population limits their ability to safeguard their spatial rights and maintain their social practices. The community inhabits a compact area, and the inclusion of young families in a while also relying on the community infrastructure increases the population availing itself of communal institutions and the prestige of the community and its leaders. The boys' school opens with the beginning of settlement, but the girls' school was opened later, and in the interim, the girls had to commute daily to Stamford Hill. As the spatial predominance of Satmar is increased, the veteran residents of Canvey Island are concerned about its impact on the public spaces and its possible ramifications.

Despite some resemblances, the organised entry into Canvey Island differs from the classic invasion–succession model (Park 1936; Wirth 1956), according to which relatively free individuals move spontaneously into areas of higher-status populations, using private capital, and also from a racial blockbusting in US cities, which involved planned invasions even if the newcomers were not the planners. In both cases, the newcomers created spatial dominance. Since these areas reflect compromises that were made between the preference for segregation and urban constraints, they tend to be heterogeneous. In Canvey Island, the group action was supported by community capital, and organised collaboration and top-down implementation. The gradient of prices on Canvey Island shows a moderate drop in prices and after that a moderate rise, in an attempt to avoid the formation of an internal market.

The groups' boundaries reflect compromises that were made between the preference for segregation and urban constraints; and for now, they tend to be more heterogeneous than the traditional Haredi enclave of Stamford Hill, for example. A direct collaboration of individuals with their leadership has gradually inserted the cultural language of Satmar into Canvey Island and reshaped the boundary between acceptable and unacceptable uses. This involves property lines and religious composition (Anderson and O'Dowd 1999), draws the boundaries between identities and provides a particular utility to the broader mission of creating a new Haredi settlement. With this regards, it seems that there is Eruv around Satmar current compound in the former Castle view school since adults carry books on Shabbat and push baby strollers and children ride on scooters. As the process is continuing, it is clear that the Satmar learned from previous experience. The properties are purchased sporadically, in a way that leads to a heterogeneous residential pattern. By imposing group discipline on people who considered to buy flat in the free market, avoiding homogeneous "ghetto" enclave, opening up and cooperating with the local media and mix-housing with the locals, Satmars can achieve a positive attitude of the veteran residents. Previous experience indicates that over time, there will be two entirely separate communities on the island, leading parallel lives. This could work very well on the basis of mutual respect and "each to their own". Despite the Satmars effort to avoid inner community market, it is anticipated that higher demand will increase properties' prices. This will improve the equity of those who leave Canvey Island, but decrease the ability of the next generation of Canvey Island's current residents to buy property there and may lead to resentment and friction.

References

Alexander ER (2002) The public interest in planning: from legitimation to substantive plan-evaluation. Plan Theory 1:226–249

Almond G, Appleby S, Sivan E (1995) Fundamentalism: genus and species. In: Marty M, Appleby S (eds) Fundamentalisms comprehended. University of Chicago Press, Chicago, pp 399–424

Altvater E (1998) Theoretical deliberations on time and space in post-socialist transformation. Reg Stud 32(7):591–605

Anderson L, O'Dowd J (1999) Borders, border regions and territoriality: contradictory meanings, changing significance. Reg Stud 33(7):593–604

Andreoni J (1995) Cooperation in public-goods experiments: kindness or confusion? Am Econ Rev 85:891–904

Aran G, Stadler N, Ben-Ari E (2008) Fundamentalism and the masculine body: the case of Jewish ultra-orthodox men in Israel. Religion 38(1):25–53

Bankston CL III, Zhou M (1995) Religious participation, ethnic identification, and adaptation of Vietnamese adolescents in an immigrant community. Sociol Q 36(3):523–534

Beyers J (2004) Voice and access. Eur Union Politics 5(2):211–240

Boal FW (2008) Territoriality on the Shankill-Falls divide: being wise after the event? Ir Geogr 41(3):329–335

Booth P (1996) Controlling development: certainty and discretion in Europe, the USA and Hong Kong. UCL Press, London

Campbell H (2002) Planning: an idea of value (inaugural lecture). Town Plan Rev 73(3):271–288

Chakak Y (2004) Vocational training for ultra-orthodox men. The Floersheimer Institute for Policy Studies, Jerusalem (in Hebrew)

Clarck TN (2000) Old & new paradigms for urban research: globalization & the fiscal austerity & urban innovation project. Urban Aff Rev 36(1):3–45

Cullingworth JB (1994) Alternate planning systems: is there anything to learn from abroad? J Am Plann Assoc 60:162–172

Dean H (2011) Welfare, identity and the life course. In: Baldock J, Mitton L, Manning N, Vickerstaff S (eds) Social policy. Oxford University Press, Oxford, UK, pp 201–244. ISBN 9780199570843

Denzau T, Munger M (1986) Legislators and interest groups: how unorganized interests get represented. Am Polit Sci Rev 80(1):89–106

Drezon-Tepler M (1990) Interest groups and political change in Israel. State University of New York Press, New York, NY

Feitelson E (2011) Issue generating assessment: bridging the gap between evaluation theory and practice? Plann Theory Pract 12(4):549–568

Flint Ashery S (2015) Public welfare or sectarianism: a new challenge for planning. Plan Theory & Pract 16(3):299–318

Flint Ashery S (2017) 'Decision not to decide': a new challenge for planning. Eur Plan Stud 25(6):1076–1098

Flint Ashery S (2018a) Vertical examination of residential patterns in Whitechapel. In: Micro-residential dynamics, pp 35–49). Springer, Cham

Flint Ashery S (2018b) Private and community housing. In: Micro-residential dynamics, pp 51–71. Springer, Cham

Flint Ashery S, Benenson I, Alfasi N (2014) Group and collective behaviour: residential patterns in Kiryat HaYovel, Megamot–Szold Institute for Behavioural Studies (Hebrew)
Funk C (2000) The dual influence of self-interest and societal interest in public opinion. Polit Res Q 53(1):37–62
Gonen A (2006) Between Torah learning and wage earning: the London experience and lessons for Israel. The Floersheimer Institute for Policy Studies, Jerusalem (in Hebrew)
Gottman J (1973) The significance of territory. University of Virginia Press, Charlottesville
Granovetter M (1978) Threshold models of collective behavior. Am J Sociol 83(6):1420–1443
Hasson S (2001) The struggle for hegemony in Jerusalem: secular and ultra-orthodox politics. The Floersheimer Institute for Policy Studies, Jerusalem (in Hebrew)
Hechter M, Kanazawa S (1997) Sociological rational choice theory. Annu Rev Sociol 23:191–214
Hendricks C (2006) When the forum meets interest politics: strategic uses of public deliberation. Politics Soc 34(4):571–602
Johnston R, Poulsen M, Forrest J (2007) The geography of ethnic residential segregation: a comparative study of five countries. Ann Assoc Am Geogr 97(4):713–738
Kolossov V (2005) Border studies: changing perspectives and theoretical approaches. Geopolitics 10(3):606–632
Kong L (2001) Mapping 'new' geographies of religion: politics and poetics in modernity. Prog Hum Geogr 25(2):211–233
Lowery D, Gray V, Fellowes M (2005) Sisyphus meets the Borg: economic scale and inequalities in interest representation. J Theor Polit 17(1):41–74
Malmberg M (1980) Human territoriality: survey of behavioural territories in man with preliminary analysis and discussion of meaning. Mouton Publishers
McKay D (1996) Urban development and civic community: a comparative analysis. Br J Polit Sci 26(1):1–23
Miller D (1999) The norm of self-interest. Am Psychol 54(12):1053–1060
Newman D (2003) On borders and power: a theoretical framework. J Borderl Stud 18:13–24
Nylund K (2001) Cultural analyses in urban theory of the 1990s. Acta Sociol 44:219–230
Oppenheim F (1975) Self-interest and public interest. Political Theory 3(3):259–276
Paasi A (1996) Territories, boundaries and consciousness. Wiley, New York
Park RE (1936) Succession, an ecological concept. Am Sociol Rev 1(2):171–179
Poppelaars C (2007) Resource exchange in urban governance: on the means that matter. Urban Aff Rev 43(1):3–27
Ratner R, Miller D (2001) The norm of self-interest and its effects on social action. J Pers Soc Psychol 81(1):5–16
Romain J (2018) Stamford Hill's ultra-orthodox Jewish community is moving to Canvey Island—as a rabbi, I'm cautiously optimistic. The independent, Tuesday, 9 Jan 2018. https://www.independent.co.uk/voices/orthodox-judaism-canvey-island-bbc-stamford-hill-integration-prejudice-a8149546.html
Sack RD (1983) Human territoriality: a theory. Ann Assoc Am Geogr 73(1):55–74
Sack RD (1986) Human territoriality: its theory and history. Cambridge University Press, Cambridge
Saegert S, Winkel GH (1990) Environment psychology. Annu Rev Psychol 41:441–477
Salisbury RH (1969) An exchange theory of interest groups. Midwest J Polit Sci 13(1):1
Sharkansky I (1996) Governing Jerusalem. Wayne State University Press, Detroit
Sprinzak E (1986) What was right in his own eyes: illegalism in Israel society in Hebrew. Library, Tel Aviv, Poalim
Stone D (2002) Policy paradox: the art of political decision making. Norton, New York
Taylor N (1998) Urban planning theory since 1945. Sage, London, UK
Tewdwr-Jones M (1999) Discretion, flexibility, and certainty in British planning: emerging ideological conflicts and inherent political tensions. J Plan Educ Res 18:244–256
Valins O (2003) Stubborn identities and the construction of socio-spatial boundaries: ultra-orthodox Jews living in contemporary Britain. Trans Inst Br Geogr 28(2):158–175
Wirth L (1956) Community life and social policy. University of Chicago Press, Chicago
Yiftachel O (2006) Ethnocracy: land and identity politics in Israel/Palestine. University of Pennsylvania Press, Philadelphia

Chapter 11
The Hasidic Communities of Stamford Hill and Canvey Island

Abstract A comparison between the residential dynamics in Stamford Hill and Canvey Island enables direct examination of a variety of non-economic segregation theories, which demonstrate that the individual's ability to implement his or her residential preferences is also affected by the group's level of community organisation and the characteristics of the surrounding urban space. Religious and communal identity appear as the driving force behind the micro-dynamics in Stamford Hill, allow relatively small Hasidic groups living close to the city centre to reproduce their physical proximity in space, and maintain their communal identity. Considering the high fertility rates and high residential demands, the council is limited in supporting the Hasidic communities' housing needs. More specifically, we could expect micro-segregation mechanisms to operate in multicultural urban hubs as a result of people's urge to feel comfortable in their residential surroundings. The housing shortage at Stamford Hill prompted the Satmar community to initiate a group action and send pioneers to settle in the peripheral area of Canvey Island. The quality of life, in terms of living conditions and housing prices, is significantly higher than in the old enclave, with a relatively low density of two persons per room, and the construction irregularities are negligible. Despite the concern of the veteran residents of the implications of the entry of newcomers on the composition of the community and house prices, the support of the local authority increases the chances of Canvey Island to become a model of harmonious standalone coexistence. The micro segregation pattern of the Satmar community in Stamford Hill and around JCoCI in Canvey Island suggests that seemingly homogeneous inner-city neighbourhoods actually include subareas created by religious, social or cultural relations.

Keywords Non-Economic segregation · Religious and communal identity · Overdensity · Group action · Community support mechanisms · Community discipline · Living conditions · Construction irregularities

A comparison between the residential dynamics in Stamford Hill and Canvey Island enables direct examination of a variety of non-economic segregation theories which demonstrate that the individual's ability to implement his or her residential preferences is also affected by the group's level of community organisation and the characteristics of the surrounding urban space.

S. Flint Ashery, *Spatial Behavior in Haredi Jewish Communities in Great Britain*, The Urban Book Series, https://doi.org/10.1007/978-3-030-25858-0_11

Religious and communal identity appear as the driving force behind the micro-dynamics in Stamford Hill, allow relatively small Hasidic groups living close to the city centre to reproduce their physical proximity in space and maintain their communal identity. The Hasidic individuals live within mixed Haredi and non-Haredi Jews, each group encompass various subcommunities. Each of these communities is having a distinct title, heritage and lifestyle characteristics. Shaping the individuals and the community characteristics against other Hasidic communities living nearby, the individuals carry the responsibility for their own Torah lifestyle and sometimes affiliated with more than one synagogue. With costly housing prices and over density of average five people per room, it appears that their *group behaviour* residential decisions are primarily directed towards living with a sufficiently large circle of "friends" who match the person's self-identity. The very existence of a few neighbours belonging to the same actual or anticipated group can create a satisfactory living environment and minimise the sense of unfamiliarity related to living among "others". Considering the high fertility rates and high residential demands, the council is limited in supporting the Hasidic communities' housing needs. As a result, and despite a high level of enforcement, there are many irregularities—e.g. additional room, shed to room, etc., in the neighbourhood. Stamford Hill-like residential dynamics are therefore expected to emerge in other enclaves. More specifically, we could expect micro-segregation mechanisms (Flint Ashery 2018) to operate in multicultural urban hubs as a result of people's urge to feel comfortable in their residential surroundings.

The housing shortage at Stamford Hill prompted the Satmar community to initiate a group action and send pioneers to settle in the peripheral area of Canvey Island. This group action, during which Satmar families moved in an organised manner to a defined area in Canvey Island, was top-down supported by community institutions, offering residential loans on reasonable terms to the newcomers. Being entitled to community support, newcomers are required to rent apartments owned by the community or purchase new apartments, usually nearby non-Jewish neighbours. A person or a family who buys or rents property at the market price, is breaking the community's discipline, and his children will not be admitted to the community institutions. Despite the distance, the mother community of Satmar Stamford Hill takes full responsibility for the Torah lifestyle of the individuals. The Satmars supervises the daily life in the new settlers, and they are required to rely on community institutions and infrastructures. Among other things, they must purchase Kosher food at the community store and live within the delineated area, at a distance of up to 20-min walk from the community centre. It should be noted that the quality of life, in terms of living conditions and housing prices, is significantly higher than in the old enclave, with a relatively low density of two persons per room, and the construction irregularities are negligible.

The rapid and organised entry of members of the Satmar community to Canvey Island raises questions about the relationship between the newcomers and the host community, given that the Haredi community is rapidly growing. In line with the theory, a lack of structural assimilation may be reflected by that despite the outward movement to the suburbs of sizeable proportions of European immigrants (in the American case), such move has been seen to take place in adjacent sectors of the city rather than in a random way. This has led to the over-representation of any one particular group within certain tracts of the city concerning their overall contribution to the urban population (Ward 1982, p. 264). Concerning the Satmars, will the host community, mostly unorganised non-Jewish individuals, continue to be cordial even when the Satmar community grows? Will the members of the Satmar community be able, and to what extent, to integrate into the local population despite Halakhic restrictions?

The micro-segregation pattern of the Satmar community in Stamford Hill and around JCoCI in Canvey Island could be relevant to other urban situations, enabling strangers to live together in dense urban areas. Inner-city areas bring about the proximate living of families and individuals of various social groups and lifestyles. The comparison between these two case studies suggests that seemingly homogeneous inner-city neighbourhoods actually include subareas created by religious, social or cultural relations. An update from April 2019 indicates that the small Satmar community in Canvey Island has expanded tenfold since it was established in 2017. The foundation's income has doubled over the past year and it is able to support its hundred families and the sixty students. Rabbi Jonathan Romain, the minister of Maidenhead Synagogue, explains in his article (Independent 2018) that the long-term outlook may be less rosy: "The Canvey Island experiment highlights the relative merits of the three routes facing all religious and ethnic minorities: isolation, which preserves identity but at the cost of social cohesion; assimilation, which leads to a total loss of one's distinctive traditions; and integration, which attempts to have a foot in both worlds without sacrificing either one". Despite the concern of the veteran residents of the implications of the entry of newcomers on the composition of the community and house prices, the support of the local authority increases the chances of Canvey Island to become a model of harmonious stand-alone coexistence.

References

Flint Ashery S (2018) Horizontal examination: micro-segregation mechanism in the diverse area of Whitechapel. In: Micro-residential dynamics, pp 25–34. Springer, Cham

Romain J (2018) Stamford Hill's ultra-orthodox Jewish community is moving to Canvey Island—as a rabbi, I'm cautiously optimistic. The independent, Tuesday, 9 Jan 2018. https://www.independent.co.uk/voices/orthodox-judaism-canvey-island-bbc-stamford-hill-integration-prejudice-a8149546.html

Ward D (1982) The ethnic ghetto in the United States: past and present. Trans Inst Br Geogr 257–275

Chapter 12
Summary and Conclusion

Abstract The present study compared the mechanisms used by members of the Litvish and the Hasidic communities in four residential areas in the UK to construct a private and public "demarcation line" between the abstract obligations and the spatial reality in which they are applied and expose value-based elements that ensure the communities' survival and resilience. Going beyond the case studies, identification of a flexible and adaptive approach to delineation-setting in the spatial reality of these four neighbourhoods broadens our knowledge about the spatial behaviour of the Haredi population, and the ramifications of changing policy on the strengthening of individual and collective identity in view of the challenge of exposure to "others". The conflicts between the sacred and the profane, between the individual and his community and the general population, between the centre and the periphery, raise the old question about the role of interventions such as planning. The "terrain of interests" in which social groups use planning methods and rights to achieve sectoral goals underlines the impotence of the liberal planning system to regulate the infrastructure and enforce planning laws in areas where interests clash. Addressing a problem by bringing various approaches and methods together, while at once combining data gathered at the most micro level, the project aims to further interject in more narrow public debates on residential relations.

Keywords Private and public demarcation line · Communal obligations and the spatial reality · Communities' resilience · Residential competition

The present study compared the mechanisms used by members of the Litvish and the Hasidic communities in four residential areas in the UK to construct a private and public "demarcation line" between the abstract obligations and the spatial reality in which they are applied. While the Hasidim are concentrated around the Admor, the Litvish have a more flexible spiritual centre. The existing literature assumes that group identity and intergroup competition in the modern city can affect religious development. This is especially true for the Haredi Jewish communities, which operate social, organisational and spatial mechanisms that create a delineation between religious obligations and the reality, to ensure the survival and resilience of the community in competition with neighbouring groups. In the context of an unstable world economy and social changes, the Haredi Jewish communal identity faces

S. Flint Ashery, *Spatial Behavior in Haredi Jewish Communities in Great Britain*,
The Urban Book Series, https://doi.org/10.1007/978-3-030-25858-0_12

daily challenges. The present study examined the residential patterns of the Litvish communities in Golders Green and Gateshead and the micro-residential dynamics of the Hasidic communities in Stamford Hill and Canvey Island and compare the mechanisms serving the communities in structuring the private and collective delineation between halakhic requirements and the reality within which they are implemented, and exposes value-based elements that ensure the communities' survival and resilience.

Within the Litvish world, the community in Golders Green organises itself while competing within diverse population groups. Individuals and families themselves bear the responsibility for building and maintaining the next generation's communal identity. This reality brings exposure to and rationalisation of the employment market and consumer culture, with a weakening of religious reliance. Personal responsibility allows the Litvish men a degree of flexibility when combining religious studies and employment, and relative freedom in adapting the communal lifestyle to his abilities and professional preferences. Elements of socialisation and choice in the life cycle—the individual's attachment to the community from infancy, his religious education, the supervision of his young family and his position as an independent adult who makes a living in order to provide an economic basis for the life cycle to continue—strengthen the individual's identity on an ongoing basis through comparison with the external society. In Gateshead, social norms and mechanisms of supervision attach the men to the Yeshiva world and prevent most of them from joining the general workforce. Dependence of a broad population of full-time scholars and their families on shrinking outside support raises the spectre of economic collapse, while competition over the allocation of resources increases internal and external conflicts as the community expand.

Within the Hasidic world, the Hasidic communities are organised in a "top-down" manner. The individual's dependence on the ongoing support of the community creates a high level of awareness of the leadership's values and expectations. This linguistic differentiation is expressed also in the urban space, where the Hasidic communities maintain separation from neighbouring communities. In Stamford Hill, they compete over space and resources with a range of other Jewish communities. The view that a Haredi person is more likely to adopt a different Jewish identity than he is to abandon his religion altogether has prompted the Hasidic leadership in Stamford Hill to adopt rigid boundaries that have led to the creation of homogeneous residential patterns. Thus, it may be the relatively rigid boundaries drawn by the Hasidic leadership in Stamford Hill that create schism both within the community and externally, in relation to the general society. The Satmar newcomers to Canvey Island operate as an organised and disciplined minority in its dealings with non-organised individuals. Although being a minority in Canvey Island, the community enjoys over-representation, allowing its appointees access to senior office-bearers and increasing their ability to influence policy-makers.

These trends, with their various effects on the Haredi identity, find expression in the residential enclaves—both in the inner city of London and in the periphery—and raises the possibility that countering processes, which make religion a choice, have opposite trends which strengthen the communal identity. On the one hand, the clustering of the Hasidic groups within the communities' enclave and the extremism of Stamford Hill's communities define the delineated area and its spatial boundaries. On the other hand, the constant encounter of the Litvish community in Golders Green with other, mainly non-Jewish groups, serves to bond the community and soften its socio-organisational-spatial delineation. It is specifically the reality of ongoing interaction, entailing constant awareness and ongoing consolidation of identity in relation to others and modernity that has the power to structure a community with a strong, positive group identity. This relative flexibility allows members of the Litvish community to adopt a more materially comfortable lifestyle while maintaining communal norms while living alongside different populations. It is possible that it is specifically these individualist modes of integration that contribute to the community's survival.

In line with the theory, the Litvish enclave in Gateshead is described as 'suburban clumping' by Rosenthal, due to what he describes as the 'one-dimensional' (occupation) only nature of the assimilation process. Similar processes started by the Hasidic community of Satmar in Canvey Island are identified by Varady and Mantel (1981) and Stern (1984), resulting in new residential concentrations in the suburbs which have become commonly known as the 'gilded ghettos' (Kramer and Leventman 1961). Based on the case studies of Golders Green and Stamford Hill, this study suggests that adopting some flexibility of the mechanisms involved in maintaining communal identity in London might strengthen the resilience of the more isolated and peripheral communities. These mechanisms could allow for a combination of work and study, and prevent abandonment of communal identity in non-funded conditions since the communal identity of those remaining within the Yeshiva framework solely because of the economic support that it offered would likely be weakened. In adopting that flexibility, attention should be paid to the ramifications of transferring responsibility for socialisation from the institutions to the family since a considerable number of earners may turn back to the home to take a significant role in this process. Such training, at the accepted standard for the British workplace, could increase income levels for all Haredi earners in the periphery, to the national average. Haredi involvement in the workforce would create day-to-day contact among different sectors. Greater equality in bearing the economic burden and the possible contribution of Haredi workers to the general economy could increase tolerance among the general public, lessen antagonism towards the Haredi sector, and facilitate funding for a cadre of gifted religious scholars.

These conflicts, between the sacred and the profane, between the individual and his community and the general population, between the centre and the periphery, raise the old question about the role of interventions such as planning. The "terrain of interests" (Flint Ashery 2015, 2017, 2019), in which social groups use planning methods and rights to achieve sectorial goals, underlines the impotence of the liberal planning system to regulate the infrastructure and enforce planning laws in areas where interests clash (Flint Ashery et al. 2014). The Haredi do not bargain but augment their potential influence on decision-making to achieve sectorial goals. When official allocations are inadequate, they may resort to unofficial modes of action. Conversely, residents of the general population regard themselves as autonomous homeowners. They expect the government to support liberal views and demand that the authorities defend their rights and stop sectarian allocation of public resources.

Going beyond the case studies, identification of a flexible and adaptive approach to delineation-setting in the spatial reality of these four neighbourhoods broadens our knowledge about the spatial behaviour of the Haredi population, and the ramifications of changing policy on the strengthening of individual and collective identity in view of the challenge of exposure to 'others'. This study contributes two new ideas to the knowledge base of residential behaviour and planning. The first is the in-depth micro-resolution analysis of real residential processes, which other studies in the field have overlooked as they have tended to work with more aggregate patterns. The second is an identification of space-culture-oriented directions within groups that will suggest an integrated understanding linking socio-religious-cultural norms and the urban environment. Addressing a problem by bringing various approaches and methods together, while at once combining data gathered at the most micro-level, the project aims to further interject in more narrow public debates on residential relations. The respective delineations exposed in these communities raise the possibility of their adoption and implementation in other enclaves. This may have particular relevance to the question of processes of integration or rejection of communities that refrain from integration in the general society. Could the bond between members of such groups and the general society be strengthened through reinforcement of their communal identity, such that they could be integrated without the threat of disintegration? Could a strengthening of group resilience and individual identity lead to decreased hostility among different groups? This study indicates that social, organisational and spatial mechanisms existing within Haredi society may exist within other communities as well and may have the power to influence society at large.

References

Flint Ashery S (2015) Public welfare or sectarianism: a new challenge for planning. Plan Theory & Pract 16(3):299–318

Flint Ashery S (2017) 'Decision not to decide': a new challenge for planning. Eur Plan Stud 25(6):1076–1098

Flint Ashery S (2019) Residential interrelationship: the case study of the Sylheti and Dhaka communities of Whitechapel. In: Kyriakidis P, Hadjimitsis D, Skarlatos D, Mansourian A (eds), Stichting AGILE. ISBN 978-90-816960-9-8

Flint Ashery S, Benenson I, Alfasi N (2014) Group and collective behaviour: residential patterns in Kiryat HaYovel, Megamot–Szold Institute for Behavioural Studies (Hebrew).

Kramer JR, Leventman S (1961) Children of the Gilded Ghetto: conflict resolutions of three generations of American Jews. Archon Books, Hamden, Connecticut 1969 [c1961]

Stern S (1984) Changes in the residential patterns of a Jewish community in the west: the case of Basel, Switzerland'. In: Proceedings Association Israeli Geographic Conference (Haifa), vol 243, p 267

Varady DP, Mantel SJ (1981) Toward an improved estimate of the size of the Jewish population of Cincinnati. Technical Report to the Jewish Federation of Cincinnati

UCL Centre for Advanced Spatial Analysis

Codes of Ethics

This research project conducted with full compliance of research ethics norms of the European Commission, The Marie Skłodowska-Curie actions 2011.

https://ec.europa.eu/programmes/horizon2020/en/h2020-section/ethics

Title of Study: Residential relations within the Haredi communities in the UK.

Name and Contact Details of the Researcher (PI):_ Dr. Arch. Shlomit Flint shlomit.flint@ucl.ac.ukc.

Name and Contact Details for Data Protection queries: data-protection@ucl.ac.uk.

This study has been approved by the UCL Research Ethics Committee: Project ID number: 13921/001.

S. Flint Ashery, *Spatial Behavior in Haredi Jewish Communities in Great Britain*, The Urban Book Series, https://doi.org/10.1007/978-3-030-25858-0

Bibliography

Abrams D, Hogg MA (1988) Comments on the motivational status of self-esteem in social identity and intergroup discrimination. Eur J Soc Psychol 18(4):317–334

Altman I, Chemers MM (1984) Culture and environment (no 2). CUP Archive

Bauman Z (2001) The individualized society. Polity, Cambridge

Baumel S (2003) Black hats and holy tongues: language and culture. Among British Haredim. Eur Judaism 36(2):91–109

Berman E, Klinov R (1997) Human capital investment and nonparticipation: evidence from sample with infinite horizons (or Jewish Father Stops Going to Work). Discussion paper series no. 97.05. The Maurice Falk Institute for Economic Research in Israel, Jerusalem

Caplan K, Stadler N (2009) Leadership and authority in Israeli Haredi society. Hakibutz Hameuhad and the Van Leer Institute, Tel Aviv

Dwyer C (1993) Constructions of Muslim identity and the contesting of power: the debate over Muslim schools. In: Constructions of race, place and nation, pp 143–159

Feldman E (2001) Catholics and Jews in twentieth-century America. University of Illinois Press, p 40

Friedlander R, Hecht R (1996) To rule Jerusalem. University of California Press, Berkeley

Godley A (2001) Jewish immigrant entrepreneurship in New York and London, 1880–1914. Palgrave, New York

Gonen A (2001) From Yeshiva to work: the American experience and lessons for Israel. Floersheimer Institute for Policy Studies, Ltd

Hakak Y (2009) Youthful bodies rebel: young men in Israeli Haredi Yeshivas today. Young 17(3):221–240

Healey P (1997) Collaborative planning: shaping places in fragmented societies. Macmillan, London

Heilman S (1992) Defenders of the faith: inside ultra-orthodox Judaism

Hogg MA, Abrams D (2003) Intergroup behavior and social identity. In: Hogg MA, Cooper J (eds) The Sage handbookof social psychology. Sage, London, pp 407–431

Hogg MA, van Knippenberg D (2003) Social identity and leadership processes in groups. In: Zanna MP (ed) Advances in experimental social psychology, vol 35. Academic Press, San Diego, CA, pp 1–52

Jaspal R, Breakwell G (2014) Identity process theory: identity, social action and social change. Cambridge University Press

Jepperson RL (1991) Institutions, institutional effects, and institutionalism. In: Powell WW, DiMaggio (eds) The new institutionalism in organizational analysis. University of Chicago Press, Chicago, pp 143–163

Jerusalem Municipality (2013) Annual educational report, Ultra-Orthodox Education

Kellner H (2003) However imperceptibly: From the historical to the sublime. PMLA 118(3):591–596

S. Flint Ashery, *Spatial Behavior in Haredi Jewish Communities in Great Britain*, The Urban Book Series, https://doi.org/10.1007/978-3-030-25858-0

Kranzler G (1961) Williamsburg: a Jewish community in transition. Philipp Feldheim, Incorporated

Kranzler G (1995) Hasidic Williamsburg: a contemporary American hasidic community. Jason Aronson, Incorporated

Leider S, Möbius MM, Rosenblat T (2009) Directed altruism and enforced reciprocity in social networks. Q J Econ 124(4):1815–1851

Mobius M, Quoc-Anh D, Rosenblat TS (2004) Social capital in social networks. Retrieved from 3 Mar 2009

Moscovici S (1973) Foreword. In: Herzlich C (ed) Health and illness: a social psychological analysis. Academic Press, London/New York

Nahmias D (2004) Local governance in Israel: policy analysis and redesign. Interdisciplinary Centre Herzelya, Israel (in Hebrew)

ONS (2012) 2011 census: item non-response, item deterministic editing and item imputation. ONS, Newport

ONS (2014) Methodology note for the 2011 area classification for output areas. ONS, London. Table LC4417

Poll S (1962\2017) The Hasidic community of Williamsburg: a study in the sociology of religion. Routledge

Poll S (1995) The charismatic leader of the Hasidic community: the zaddik, the Rebbe. In: New world Hasidim: ethnographic studies of Hasidic Jews in America. State University of New York Press, Albany, pp 257–275

Sarre P (ed) Human geography today. Polity Press, Cambridge

Shaffir W (1987) Separation from the mainstream in Canada: the Hasidic community of tash. Jew J Sociol 29(1):19–35

Shaffir W (1995) Boundaries and self-presentation among the Hasidim: a study in identity maintenance. In: New world Hasidim: ethnographic studies of Hasidic Jews in America, pp 31–68

Shilhav Y (1983) Communal conflict in Jerusalem: the spread of ultra-Orthodox neighbourhoods. In: Pluralism and political geography, people, territory and state, p 100

Shilhav Y (1984) Spatial strategies of the "haredi" population in Jerusalem. Socio-Econ Plan Sci 18(6):411–418

Shilhav Y (1998) Ultra-orthodoxy in urban governance in Israel. The Floersheimer institute for policy studies, Jerusalem

Sigurdson R (2000) Crossing borders: immigration, citizenship and the challenge to nationality. In: Pratt M, Brown J (eds) Borderlands under stress. Kluwer Law International, London, pp 141–162

Silber M, Werthheimer J (1992) The emergence of ultra orthodoxy. The Uses of Tradition. Jewish Theological Seminary of America, New York, pp 23–84

Sklare M (1982) On the preparation of sociology of American Jewry. Underst Am Jewry, 261–271

Sklare M (1990) Religion and ethnicity in the American Jewish community. Soc Found Judaism, 135–145

Turner JC, Bourhis RY (1996) Social identity, interdependence and the social group: a reply to Rabbie et al. In: Robinson WP (ed) Social groups and identities: developing the legacy of Henri Tajfel. Butterworth-Heinemann, Oxford, UK, pp 25–63

Valins O, Kosmin B, Goldberg J (2001) The future of Jewish schooling in the United Kingdom: a strategic assessment of a faith-based provision of primary and secondary school education. Institute for Jewish Policy Research

Vorspan M, Gartner LP (1970) History of the Jews of Los Angeles. Huntington Library

Printed by Printforce, the Netherlands